# Research Reports ESPRIT

T0250836

Subseries PDT (Product Data Technoi. ﹍,,

Project 2010 · NEUTRABAS

Edited in cooperation with the European Commission and the
Product Data Technology Advisory Group (PDTAG)

# Springer
*Berlin*
*Heidelberg*
*New York*
*Barcelona*
*Budapest*
*Hong Kong*
*London*
*Milan*
*Paris*
*Tokyo*

H. Nowacki  (Ed.)

# NEUTRABAS

## A Neutral Product Definition Database for Large Multifunctional Systems

With contributions by
Francisco Fernández-González
Markus Lehne
Kenneth J. MacCallum
Sabine Müllenbach
Bernard Nizéry
Horst Nowacki
George R. Stephenson

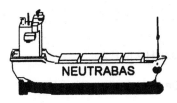

NEUTRABAS

Springer

**Volume Editor**

Horst Nowacki
Technische Universität Berlin, Fachbereich 10
Verkehrswesen und Angewandte Mechanik
Institut für Schiffs- und Meerestechnik
Salzufer 17–19, D-10587 Berlin, Germany

ESPRIT Project 2010 "A Neutral Product Definition Database for Large Multifunctional Systems (NEUTRABAS)" belongs to the Subprogramme "Computer Integrated Manufacturing" of ESPRIT, the European Specific Programme for Research and Development in Information Technology supported by the European Commission.
NEUTRABAS belongs to the second generation of ESPRIT projects (within ESPRIT II) that have built on the results of the CAD*I Project (ESPRIT Project 322) and have made significant contributions to the development of the international standard STEP (ISO 10303) for neutral product data definition and exchange. During the NEUTRABAS project the STEP methodology for neutral product modeling was still evolving so that it was necessary to extend and influence the standard for purposes of the project. NEUTRABAS has concentrated its contribution onto the areas of multifunctional product modeling and the sharing data by means of neutral, distributed databases. It has gained valuable experiences with prototype implementations of this new technology.

CR Subject Classification (1991): J.6, J.2, H.2.8, E.2

ISBN 3-540-59300-4 Springer-Verlag Berlin Heidelberg New York

CIP data applied for

Publication No. EUR 16258 EN of the
European Commission,
Dissemination of Scientific and Technical Knowledge Unit,
Directorate-General Telecommunications, Information Market
and Exploitation of Research,
Luxembourg.

Typesetting: Camera-ready by the editors
SPIN: 101486062        45/3142-543210 – Printed on acid-free paper

# Foreword

Product Data Technology encompasses the information related to all stages in the product life cycle from product design via production planning, production processes, etc., through the delivery and operation of the technical product. Product Data Technology takes a coherent, unified view of the information captured in this whole life cycle and provides methodologies to support this integrated perspective.

The Product Data Technology Advisory Group (PDTAG), which was founded in 1991 as a Special Interest Group under the auspices of the CIM Europe Division of DG III (Industry) of the Commission of the European Communities, has in 1994 become an Accompanying Measure Project in ESPRIT (Project No. 9049). PDTAG in its coordinating responsibility for European PDT developments has encouraged the formation of a new subseries on Product Data Technology within Springer Verlag's existing series of Research Reports ESPRIT. This subseries will serve as a depository for the important contributions made by ESPRIT projects to the evolving area of Product Data Technology. PDTAG is grateful to Springer for having established this subseries.

The current volume presents the results of the NEUTRABAS Project (ESPRIT 2010), dealing with the development of a neutral database for large multifunctional systems, in particular for ships and their multitudinous, complex subsystems. NEUTRABAS was the first European project aiming at an international standard based on ISO 10303 (STEP) methodology to define a comprehensive information model for ships and similar products of complex functionality, which will serve for the exchange and long term storage of product information. NEUTRABAS provided major contributions to the first generation of shipbuilding product models and gained first experiences in implementing large, complex databases exploiting and extending the new technology of the still evolving standard STEP. These developments have laid many conceptual foundations for the shipbuilding product models which are maturing as ISO standards today. The present volume documents the original concepts developed and the experiences gained by NEUTRABAS.

The editor of this volume gratefully acknowledges the excellent work and valuable assistance provided by Ms. Andrea Walter who was responsible for the preparation, layout and quality assurance of the manuscript text and illustrations.

Berlin, March 1995 — Horst Nowacki
Chairman of PDTAG-AM

# Preface

The book I have the pleasure to present here is the result of a European project developed in the frame of the ESPRIT programme of the European Commission (ESPRIT = European Specific Programme for Research and Development in Information Technology) by a team of fourteen partners including three shipyards, two research centres, five information technology companies, and four universities, from four countries, France, Germany, Spain, and the United Kingdom.

The project was aimed at the development of a neutral product definition database for large multifunctional systems. In NEUTRABAS, ships are regarded as a representative example and ideal test case for large multifunctional systems. The multifunctional aspects create special requirements for the methodology of product information modelling and database design. The objective of NEUTRABAS was to define a methodology and to demonstrate the feasibility of a standardized product description supporting a neutral data exchange and archiving capability.

The main areas dealt with by the project team were:

- Modelling methodology in compliance with the evolving international ISO Standard STEP (ISO 10303) based on the use of the modelling language EXPRESS (ISO 10303-11);
- Creation of a neutral database corresponding to the EXPRESS product model;
- Data management to exchange data between heterogeneous application systems through the neutral database and corresponding software;
- Implementation of a prototype NEUTRABAS system, and validation testing.

The NEUTRABAS project is among the first developments world-wide, only preceded and accompanied by the American NIDDESC project, to aim at an international standard for data exchanges in shipbuilding and other maritime industries. NEUTRABAS in open collaboration with NIDDESC has contributed its fair share of ideas and results to this effort. NEUTRABAS has also evaluated its experience with regard to the suitability of STEP and EXPRESS for its modelling tasks.

The present book describes the main developments and conclusions of the project. The successful issue of the project is the result of the joint efforts of all the participants. I am pleased to thank and congratulate them for their active participation.

I also take the opportunity to address my thanks to the European Commission for its efficient support of the NEUTRABAS project, and especially to the Project Officer, Mr. Jacques Bus, for his very much appreciated cooperation.

March 1995                                                                   B. Nizery
                                                              NEUTRABAS Project Manager

# Contents

# 1 Introduction

The ESPRIT II Project No. 2010, NEUTRABAS, carried out between 1989 and 1992, was aimed at the development of a neutral product definition database for large, multifunctional systems, in particular for ships. Many more complex engineering systems serve several different functions simultaneously or sequentially and this may also hold for their constituent subsystems and parts. In NEUTRABAS, ships and other maritime products are regarded as representative examples for large multifunctional systems and their product models. In addition these products typically are of a one-off production type.

It is no small ambition to capture all attributes of such products with all functional relationships which may be relevant at any stage of the life cycle in a single coherent database. NEUTRABAS can be regarded as an early pilot project with these long-range objectives.

The basis for all developments was the evolving international product data definition standard STEP (ISO DIS 10303) and its modelling language EXPRESS. The objective was to create a neutral product description based on a formal definition in STEP which could serve as standardized format for database definition, data exchange and archiving for applications in the maritime industry. One major difficulty was that the STEP standard with its generic product modelling resources, e. g. for geometry, topology, materials etc., was far from complete when the project started and until today has not fully matured although it is now approaching the IS approval stage for its initial release, which will comprise all essential modelling resource parts. Similarly the modelling language EXPRESS still underwent several changes during the NEUTRABAS development. Nevertheless it was possible for NEUTRABAS to achieve its tasks in application modelling and database development, provisionally using the available STEP methodology.

The benefits of providing standardized product databases supporting neutral data exchange and archiving capability are well known and apply also to the maritime industry.

They include:

- Improved communications within enterprises and with outside partners by linking heterogeneous systems
- Integration throughout the CIM cycle and perhaps life cycle, based on a single, coherent product model
- Distribution, decentralization, and coordination of activities within concurrent engineering environments
- Collaboration with other industries via open software bridges

The methodology of the ISO standard 10303, STEP, was designed with these strategic objectives in mind.

The NEUTRABAS Project was performed by a consortium of 14 European partners under the project leadership of the Institut de Recherches de la Construction Navale, Paris. The consortium included partners from the maritime industry, system vendors, software development firms, universities and research institutes. The partners who completed the project were:

Bremer Institut für Betriebstechnik und Angewandte Arbeitswissenschaft (D)
Bremer Vulkan AG (D)
Chantiers de l'Atlantique, Alsthom (F)
COTEC Computing Services (UK)
Decision International (F)
Digital Equipment International GmbH (D)
Howaldtswerke Deutsche Werft AG (D)
Institut de Recherches de la Construction Navale (F)
Schiffko GmbH (D)
SENER - Sistemas Marinos S.A. (E)
Technische Universität Berlin (D)
Universidad Politécnica de Madrid (E)
University of Strathclyde (UK)

The NEUTRABAS project has benefited much from its collaboration with the American NIDDESC project. This project was based on a similar initiative taken by the shipbuilding industry and the Navy in the United States. NIDDESC whose beginning preceded the start of NEUTRABAS by about two years has made many valuable inputs to ISO STEP standardization [1], [2] and is continuing to do so. NEUTRABAS has been able to build on this work and to extend the scope of the maritime product models in close liaison with the Americans. In ESPRIT III a successor project to NEUTRABAS is MARITIME (Project No. 6041).

This book gives an overview of the results achieved and experiences gained in the NEUTRABAS project. These results and experiences are primarily relevant to product data modelling based on STEP methodology for a complex multifunctional product, the ship, and to database software development for STEP. The book will also describe the approach taken toward integrating the specialized topical partial models of the ship into a general appreciation reference framework, which was developed in NEUTRABAS. It will also review the pilot implementation of a STEP toolkit for a relational STEP database implementation. Some conclusions are drawn also for necessary improvements in STEP and EXPRESS.

## 2 The NEUTRABAS Approach

### 2.1 Requirements

The goal of the NEUTRABAS project was the development of a neutral product definition database for ships and similar large multifunctional products. This necessitated comprehensive developments with regard to both product model definition and database realization. The project thus encompassed the following principal tasks:

- Formal definition of a ship product information model.

- Specification of a format for the realization of a neutral database, which may be based on relational database technology (ANSI Standard SQL) or on evolving object-oriented database systems.

- Implementation of the neutral database for pilot applications in shipbuilding.

- Development of prototype pre- and postprocessors for a neutral data exchange between existing shipbuilding software systems via the neutral database.

These objectives were to be achieved by using a neutral product data technology approach based on STEP. The required properties for the neutral database were primarily the following:

- Neutral product definition method, using the standardized methodology of STEP and its modelling language EXPRESS (ISO 10303-11, available in Committee Draft form [3]).

- Completeness in scope, encompassing all relevant product attributes throughout the life cycle for the range of applications supported..

- Flexible structure, i. e., selective access to the database from multiple functional viewpoints.

- Open architecture, suitable for use in distributed systems via open systems interconnection links (OSI networks).

These requirements were of particular relevance for the modelling approach. They necessitated a very systematic discipline in defining the model, much more so than for single-purpose databases. The emphasis was placed on flexibility, openness and multifunctionality of the database. The product model itself had to have a corresponding structure.

### 2.2 Approach

In response to these requirements the following modelling approach was taken in NEUTRABAS:

- Information analysis of maritime products, in particular ships, using STEP principles and modelling resources.

- Development of ship partial product models, primarily for ship structural systems, spatial arrangements and outfitting systems. Each partial model is formulated as an independent conceptual schema. The schemas are linked primarily by sharing resources. The partial product models are forming a non-redundant set of conceptual schemas whose union is the complete ship product model. The current set of partial models is of course still incomplete and will have to be supplemented by many other partial models before a comprehensive coverage of all shipbuilding applications will be achieved.

- The partial product models were documented initially by NIAM diagrams. NIAM is the acronym for Nijssen Information Analysis Method [4]. It is a formal method for the definition of semantic information models. It uses a notation consisting of graphical symbols for objects, roles (binary relationships) between objects, and constraints on objects and roles. Information models described by NIAM diagrams result in intuitively readily comprehensible representations of the model so that domain experts can analyze, discuss and criticize them without having to be experienced modellers. This is why NIAM was chosen by NEUTRABAS as the initial documentation format for the product models. It supported a natural dialogue between domain and modelling experts.

Details about the NIAM notation, which is used extensively in this book, are found in Annex V.

- The great majority of the NEUTRABAS models were later also documented more precisely in EXPRESS and its graphical version EXPRESS-G. Unfortunately the EXPRESS language was still continually changing during the NEUTRABAS development. Therefore when the project reached its database implementation stage it had to base its product models on the EXPRESS version then available [3]. For most models only minor syntactical updates would be required to bring them into conformance with the final version of EXPRESS (ISO IS 10303-11), which is in effect now.

- For integration the developed partial product models were subsequently examined for consistency, non-redundancy and required integration points. A framework architecture was proposed in the end of the project into which the existing schemas can be embedded after some pruning. This reference framework for ship applications provides an instrument of top-down integration for existing and future partial models.

- It is possible to select from the pool of partial models and their resources those subsets of entities which are shared by certain application areas. These subsets can then be organized as self-contained schemas in the manner of Application Reference Models within STEP Application Protocols. This is the approach adopted by NEUTRABAS for defining its prototype product models used as mockups for the implementation of demonstration systems. A more official policy in STEP for defining Application Protocols and implementation subsets thereof emerged only gradually toward the end of NEUTRABAS [5]. It could therefore not be followed by NEUTRABAS and is still not fully defined.

NEUTRABAS followed this approach as far as possible in the given time-frame. A complete top-down integration of the product model based on the recommended framework was not yet achieved. Further work at ISO level is now in progress.

## 2.3  NEUTRABAS Product Models

An overview of the general structure of the NEUTRABAS product model as it was originally organized in the beginning of the project is presented in Fig. 1. It shows an overall Ship Information Model which consists of the following Partial Models:

> Global Information Model
> Spatial Information Model
> Structural Information Model
> Outfitting Information Model

These Partial Models are self-contained models and can be described by independent EXPRESS schemas. They are only loosely coupled from this overall perspective as the figure implies. They nevertheless share certain concepts by reference to models or submodels for spaces, moulded surfaces and other geometry, coordinate systems etc. Many of the shared concepts correspond to or can be mapped onto entities available at the level of STEP Integrated Generic Resources (STEP Parts 41 through 49). The application specific concepts for the ship product model are contained in the four Partial Models mentioned above. The following subsections describe these models in overview. Greater details are given in Chapter 3.

### 2.3.1  The Global Information Model

This Partial Model comprises all those properties which are suitable to characterize the ship and its functional capabilities in total. This information may be relevant to any application context or life cycle stage. It pertains to the whole ship as a product rather than to any of its subsystems and components. It may still include an overview of the functional capabilities available in a certain ship.

This model was later developed in a somewhat more limited scope and therefore re-named into Ship Principal Characteristics Model, which is described in Section 3.1.

### 2.3.2  The Spatial Information Model

This model is concerned with the representation of the form and structure of infor-mation associated with the internal spatial organisation and topology of the ship-building product. The model covers aspects such as the representation and organisation of the major internal and external surfaces including their geometrical and topological characteristics, and the representation and organisation of the resul-ting internal enclosed spaces and compartments. Implicit in the model is the concept of reference systems, both on a global and a local basis, which provide the means for positioning and orienting spatial and structural items within the overall product model.

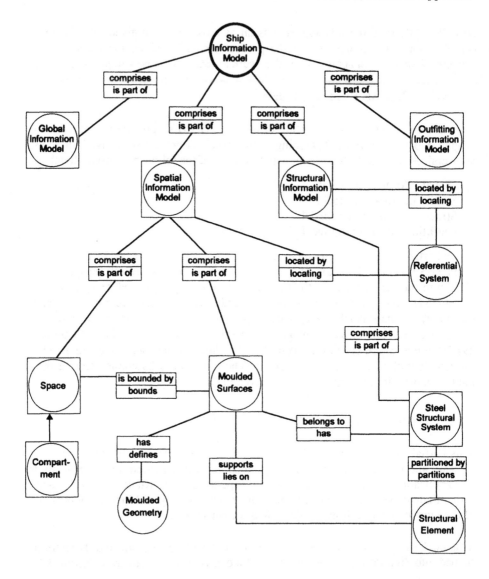

**Fig. 1** Structure of Ship Information Model with Reference to Partial Models, especially the Spatial Information Model and Structural Information Model (NIAM Diagram)

This spatial organisation model reflects the evolution of the product definition through the various life-cycle stages and allows different views of the information which support other partial models of the product such as the Structural Information Model and the Outfitting Information Model. The aim of this spatial information model is to support the many applications which require space-oriented product information pertaining to the various life-cycle stages. In order to realize this complete, full life-cycle spatial information model, the formal specification for it includes:

- Higher level entities (complex constructs) such as:
  Spaces, zones, references, types, boundaries, requirements and functions;
- Lower level entities (basic constructs) such as:
  Coordinates, dimensions, characteristics, properties and identifications.

More details are given in Section 3.3.

### 2.3.3 The Structural Information Model

The purpose of this model is to define completely the information needed in ship-building for data exchange concerning the ship hull structures.

Furthermore, the Structural Information Model adds to the Hull Spatial Organisation Model the necessary data to support the integration of the Outfitting Information Model within the higher level Ship Information Model.

This model defines the structure of the product as it is required throughout its life-cycle, from design through production engineering and manufacturing and through the operation of the ship. Geometry and topology are information categories which are conceived during the design stages, while other physical relationships are present during the production, outfitting and operation stages. Therefore, specification for different views is included covering:

- Geometry, topology, interactions, materials, connectivity, assemblies and decompositions, functions, standards, activities, legal and managerial aspects, fabrication etc.

The information included in this model concerns all structural components which belong to the ship structure including:

- Hull, substructures, assemblies, structural elements, primary items, plates, shapes, openings, stiffeners, joints, connections, prefabricated units, calculation units, outfitting units etc.

More details are given in Section 3.4.

### 2.3.4 The Outfitting Information Model

The purpose of this model is the definition of the ship's outfitting systems from a physical and functional viewpoint. From a physical viewpoint these systems and their components are characterized by their shape, location, material etc. like other product parts and assemblies. From a functional viewpoint each system is described by one or several functions which it performs according to its single or multiple roles in the operation of the ship. In addition this model describes certain administrative properties of the outfitting systems with regard to the process in which the systems are involved (schedule, time and cost).

### 2.3.5 Integration of the Models

The Partial Models described in the preceding subsections were developed rather
independently during NEUTRABAS in order to keep each model selfcontained.
Therefore a major overlap among these models, which shared many concepts, was
inevitable. During the final stage of the NEUTRABAS project an integration of these
models was therefore undertaken in order to remove all these redundancies and to
merge the models into one coherent set of schemas. A new schema architecture con-
forming with STEP guidelines and EXPRESS rules was developed and presented in
NEUTRABAS Deliverable 1.2.6 [6].

The resulting general NEUTRABAS schema architecture is shown in Fig. 2. This
new schema architecture consists of four main levels:

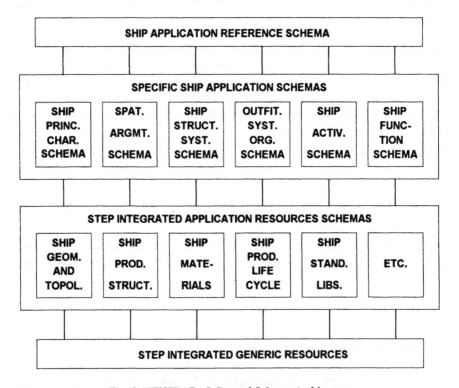

**Fig. 2** NEUTRABAS General Schema Architecture

- **Ship Application Reference Model**
  A high-level model capturing the information content and structure of the product
  from a generalized viewpoint so that any specific application view is a special case
  of this reference model. The model is thus intended to provide a unifying conceptual basis for many different applications. It consists of two schemas which are
  described in more detail in Chapter 4.

- **Specific Ship Application Schemas**
A new proposed, non-redundant schema structure for reorganizing the NEUTRABAS Partial Models into a coherent set of application specific schemas. Currently the following schemas are comprised in this set:

  - Ship Principal Characteristics Schema
  - Spatial Arrangements Schema
  - Ship Structural Systems Schema
  - Outfitting Systems Organization Schema
  - Ship Activities Schema
  - Ship Function Schema

- **Ship Integrated Application Resource Schemas**
Application oriented schemas which are shared between two or more application specific topic areas. For example:

  - Ship geometry and topology (moulded ship form)
  - Ship product structure
  - Ship materials
  - Ship product life cycle
  - Ship standard part libraries

- **STEP Integrated Generic Resources**
These resources in STEP form the foundation for the neutral modelling for products by providing a pool of entities and concepts from which Application Protocols can be built. The following resource parts of STEP are relevant to NEUTRABAS:

  Part 41: Fundamentals of Product Description and Support
  Part 42: Geometric and Topological Representation
  Part 43: Representation Structures
  Part 44: Product Structure Configuration
  Part 45: Materials

Other resource parts may become pertinent when they are released as DIS or IS.


## 2.4 Multifunctional Aspects

A product function is a role that a product performs in any given context or from any assumed viewpoint. Primary functions are those for which the product is designed or destined (purpose roles). Secondary functions are related to any other kind of functional behaviour of the product in some context or environment (behaviour roles). The distinction between purpose and behaviour is not always strict. The definition of functions depends on the viewpoint of product evaluation. Similarly there is no strict limit to the number of functions a product may perform.

Consequently it is necessary to stipulate which functions will be included in a product model. NEUTRABAS has chosen a finite, though extensible set of functions to be taken into consideration. The selection of functions is based on the tasks to be performed by the database, hence the modelling approach is driven by tasks (roles) of the product (task driven product model).

A set of objects, i. e., products or components, performing a function together is called a system. Components and products may belong to more than one system. System scope and structure must be identified in the product model. The role of each component in the system must also be defined.

A product is multifunctional if more than one function is performed by the product or any of its components. This results in multiple roles or viewpoints from which the product can be evaluated. In NEUTRABAS each function is associated with a system and can be regarded separately. The approach is taken that distinct functionalities will be mapped onto multiple views of the database. Since components can be associated with more than one function and hence system, the database must support multiple streams of attribute inheritance, at least one for each function.

In practice this complex modelling task in NEUTRABAS was approached on the basis of an object-oriented methodology. This facilitated a division of labour and a step by step procedure. The complex product was broken down into physical and functional subsets, i. e., components and systems. Partial product models can be independently defined for each subset. The partial models are later integrated into a coherent product model. To facilitate the integration task, which is of course not trivial, it is a practical necessity to develop a perspective of the scope and structure of the entire product model from the beginning.

# 3 Shipbuilding Product Models

## 3.1 Overview

### 3.1.1 Scope

According to STEP methodology a global view of the intended scope of a product model is taken initially by defining the "universe of discourse" by means of a STEP Planning Model. This method describes the general scope of the product model by enumerating those properties of the product in definition space which are to be included in the model. Moreover the concepts by which the product is to be represented in terms of its geometric and topological primitives and is to be presented in terms of a medium for display are also stated in the Planning Model. This approach was also adopted by NEUTRABAS. The NEUTRABAS Planning Model is shown in Fig. 3.

On the basis of this description given in the STEP Planning Model the intended scope of the NEUTRABAS shipbuilding product models can be characterized by three main descriptors:

- **Aggregation/Decomposition**
  Any level of aggregation or decomposition needed between the levels of elementary features and parts, subassemblies and complex assemblies, and systems comprising the ship in total.

- **Characterization**
  All major properties and functions of shipboard systems and components, in particular shape, material, function, cost, strength and stability.

- **Life Cycle**
  Most, but not all life cycle stages of the ship as a product ranging from conceptual design through all intermediate stages to ship production and delivery.

It should be noted that the word system in the context of the aggregation/ decomposition descriptor denotes a set of assemblies of which the system is physically composed. In shipbuilding this interpretation is valid for the ship structural system in total and the subassemblies from which the ship is erected. However, in a multifunctional product like a ship many components play multiple roles in several systems. In modern shipbuilding a high degree of pre-fabrication and pre-outfitting is applied so that certain outfitting components are associated with a steel structure subassembly although they belong to addifferent functional system. In modelling functional systems the aggregation/decomposition should follow functional associations rather than physical ones so that the descriptors system/subsystem/ element are more appropriate in this context than system/assembly/part.

The scope of a product model in STEP is usually further detailed for its range of applications by the method of Application Activity Models, described in the Guidelines for Application Protocols [5]. In NEUTRABAS this technique was applied only to the Outfitting Systems Model, for which the results will be presented in Section 3.5. Details about the application range of the other partial models developed in NEUTRABAS are given verbally in the following subsection.

### 3.1.2 Organization

In NEUTRABAS, four distinct information models were developed to represent different aspects of the design and manufacturing stages of complex maritime structures typically represented by ships. The four major areas addressed by NEUTRABAS were associated with the following models:

- Ship Principal Characteristics Model
- Hull Spatial Organization Model
- Ship Structural System Model,
- Ship Outfitting Systems Model.

In order to complete these information models within a realistic project time scale, it was necessary to develop the four models in parallel. For that reason, it has also been necessary to develop a strategy for integrating the existing partial product models of NEUTRABAS.

Initially most NEUTRABAS product models were specified using the notation of NIAM, The Nijssen Information Analysis Mathod [4]. See Section 3.1.3 for details and Annex V for an overview of the notation.

### 3.1.2.1 Ship Principal Characteristics Model

The Ship Principal Characteristics Model was developed in Workpackage 1 of NEUTRABAS and presented in Deliverable 1.2.6 (see Fig. 11)

The principal characteristics of ships are those properties which are suitable to characterize the ship and its functional capabilities in total from an overall viewpoint. Their relevance is not limited to any particular application area or life cycle stage. They can be pertinent as general information to any application context. In shipbuilding practice shipyards often prepare ship principal data sheets to document these principal characteristics for convenient reference. Practice varies, of course, as to what will be included in such documents. In the present draft for a Ship Principal Characteristics Model it was attempted to include a broad variety of such information without trying to be exhaustive.

Engineering applications which may read or write principal characteristics information include in particular: Hull form definition, hydrostatics, stability, trim, weights and centroids, freeboard, tonnage measurement, propulsion powering, hull and propeller performance, cargo hold and tank capacities, container capacities, etc.

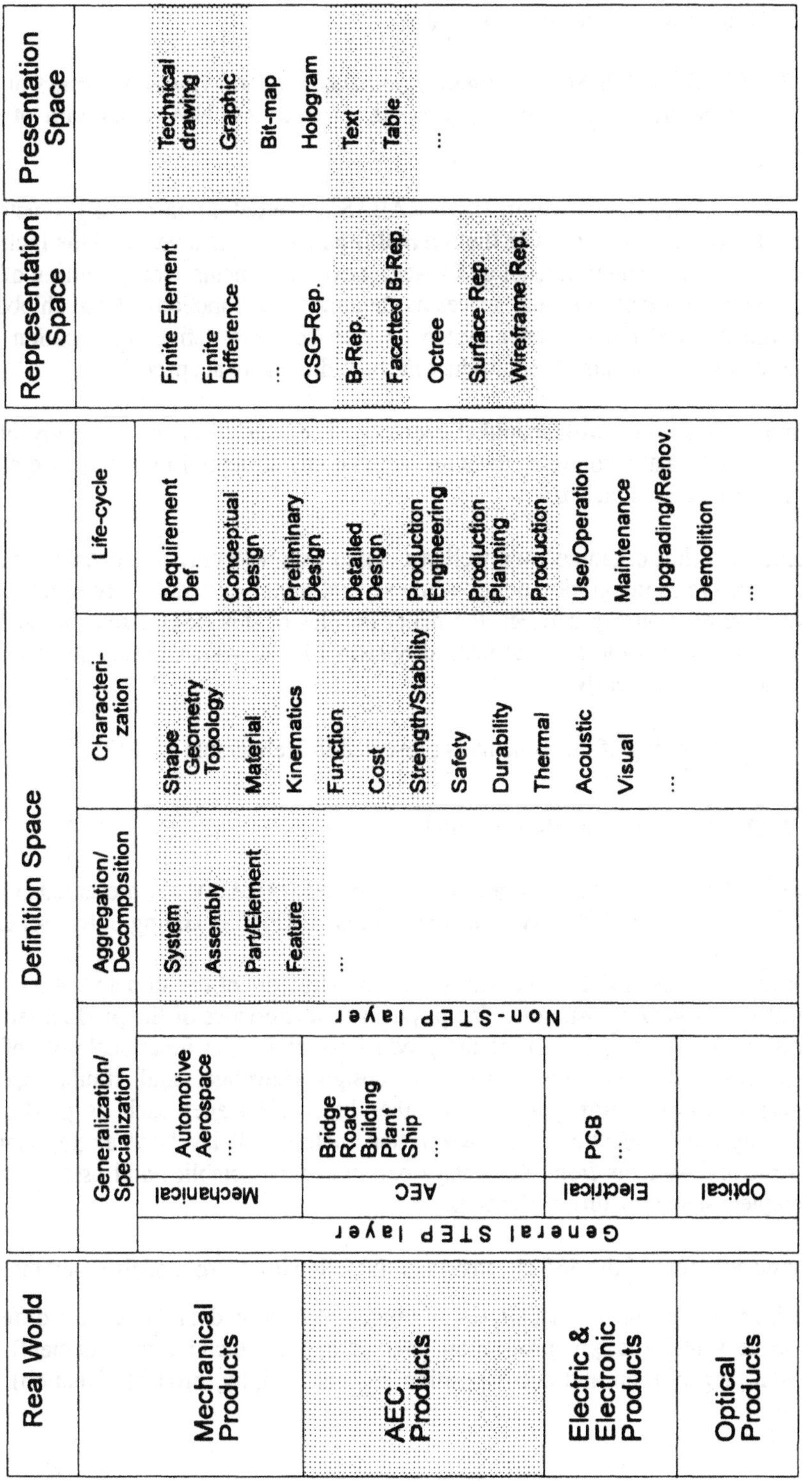

**Fig. 3** STEP Planning Model Applied to NEUTRABAS

### 3.1.2.2 Hull Spatial Organisation Model

The NEUTRABAS Hull Spatial Organization Model (HSOM) was described in full detail in Deliverable 3.2.1. Fig. 39 presents an overview NIAM diagram of this model.

The purpose of this model within NEUTRABAS is the definition of the geometrical and topological characteristics of the internal organization of a vessel. This includes the definition of the internal enclosed spaces or compartments, the definition of the surfaces by which these internal spaces are bounded, the functions of the shipboard compartments, and the reference systems which serve to define the location and orientation of components of the product in three-dimensional space.

This model deals only with moulded geometry, i. e., the idealized geometry of all reference surfaces and curves of the ship, without any structural members like plates and stiffeners being attached.

The schema is thus organized in a well-structured way with only a few major entities which can serve as integration points while several minor ones do not complicate the pattern. The relationship between the major entities of this HSOM schema and the resources as well as the Ship Structural System Model, which shares some of the same resources, are clearly evident.

An EXPRESS coding of this schema is presented in Deliverable 3.2.1[9].

### 3.1.2.3 Ship Structural Systems Model

The Ship Structural Systems Model (SSSM) of NEUTRABAS is presented in full detail in Deliverable 3.2.2, [9] which contains also EXPRESS coding of this schema.

The purpose of this model is to describe the steel structure of the ship and of its components from several viewpoints: The physical characteristics of the product (shape, material etc.), the organisation of the product structure, the functional role of the steel members and assemblies from the design viewpoint, and similar special attributes of the structure pertinent to other life cycle stages such as production engineering and production. This description includes all levels of aggregation of steel structural systems from elementary parts via subassemblies and assemblies to the complete steel structure of the ship.

The basic structure of the SSSM consists of the following main models or submodels:

• The *High Level Model* concentrates on product definition data and covers different life cycle stages (design, production engineering, production, management) and different aggregation and decomposition aspects (activity, product, function) (see Fig. 4)

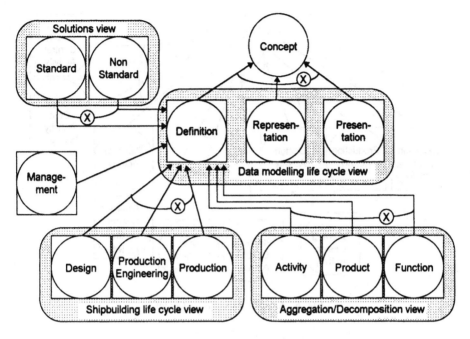

**Fig. 4** High Level Model

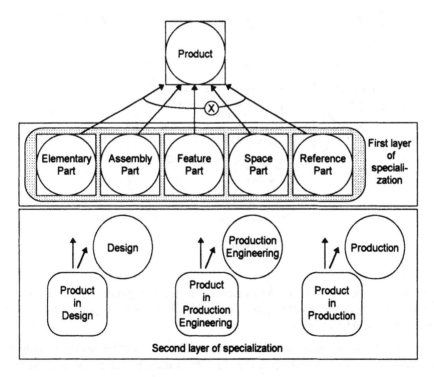

**Fig. 5** Product Information Model

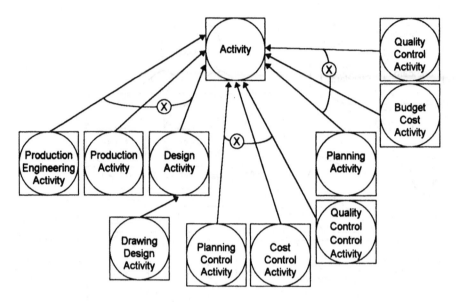

**Fig. 6**   Activity Information Model

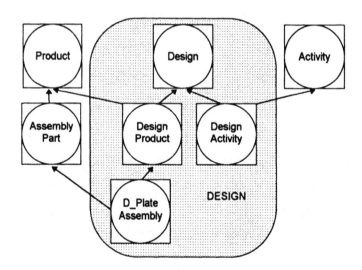

**Fig. 7**   Design Information Model

- The *Product Information Model* describes the product physical structure and its component parts (elementary parts, assembly parts, feature parts, space parts, reference parts) (see Fig. 5)

- The *Activity Information Model* refers to different activities accompanying the product during its life cycle (see Fig. 6). These include design, production engineering, production, planning, cost estimating, budgeting, quality control, cost control, etc.

The *Design Information Model* concerns the product during the design stage and encompasses the aspects of design definition, design product, and design activity (Fig. 7).

• The *Production Engineering Information Model* comprises the aspects of definition, product and activity at the production engineering stage (this model is analogous to the Design Information Model).

• The *Production Information Model* covers corresponding aspects as the previous ones (definition, product, activity) at the production stage (it is analogous to the Design Information Model).

• The *Management Model* describes various definitions for use by management activities (see Fig. 8).

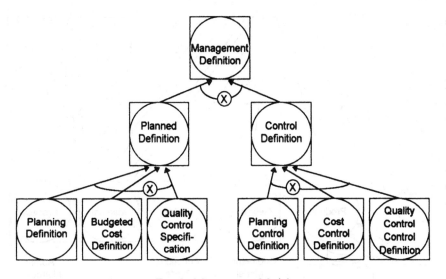

**Fig. 8** Management Model

• The *Resource Model* makes reference to STEP Generic Resources as geometry, topology and material, but also fundamentals of product description and support (Part 41) and product structure configuration (Part 44 of STEP).

The SSSM is one large coherent set of entities with a structure of one High Level Model, six application topical models, and one resource model, which facilitates a reorganization into schemas (see Section 4.2.3).

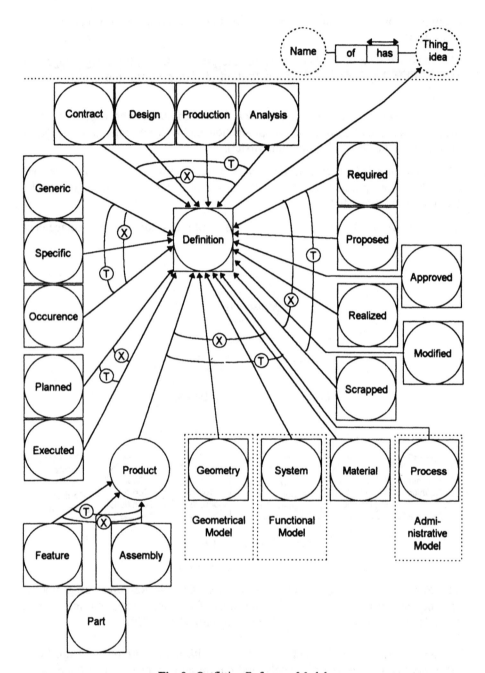

**Fig. 9** Outfitting Reference Model

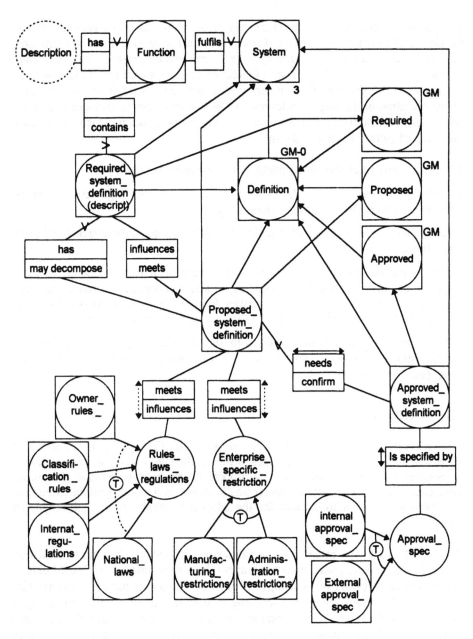

**Fig. 10** Functional Model

### 3.1.2.4 Ship Outfitting Systems Model

The NEUTRABAS Ship Outfitting Systems Model (SOSM) is presented in more detail in Section 3.5.

The purpose of this model is the definition of the ship's outfitting systems from a physical and functional viewpoint. From a physical viewpoint these systems and their components are characterized by their shape, location, material etc. like other product parts and assemblies. From a functional viewpoint each system is described by one or several functions which it performs according to its single or multiple roles in the operation of the ship. In addition, the SOSM describes certain administrative properties of the outfitting systems with regard to the processes in which the systems are involved (schedule, time and cost).

The model consists of four independent, but connected schemas:

- *Outfitting Reference Model* (see Fig. 9) which is another high level reference model accounting for outfitting system structure;

- *Functional Model* (see Fig. 10) which is related to the outfitting systems and their functions;

- *Topological and Geometrical Model* which deals with the association between system components and their geometrical and topological reference elements within the ship. This model uses entities from the Functional Model, the Ship Structural Systems Model , and the STEP Generic Resources (Part 42);

- *Administrative Model* which deals with the description of processes from an administrative viewpoint in terms of their schedule, duration, activity sequence, etc. (see Fig. 53)

### 3.1.3 Specification

The NEUTRABAS information models were generally first developed and specified by NIAM diagrams [4]. The advantages of this methodology were already mentioned in Section 2.2. An accelerated review process for the models between the author (modeller) and reader (domain expert) was achieved as a result. Certain extensions to NIAM were useful to account for the advanced complexity of multifunctional models.

NIAM belongs to the category of semantically irreducible information modelling methods and its basic paradigm is the binary relationship model. I. e., NIAM objects are associated by binary relationships called roles. The binary nature of this association limits the complexity of the resulting model, at least in comparison to object oriented models whose object relationships are arbitrarily structured and hence potentially of high complexity. Owing to the binary relationship mechanisms in NIAM models and to their semantically irreducible orientation it poses no special difficulty to use a NIAM model for different application views and to access exactly what is needed for each view.

An overview of the NIAM symbolic graphical notation, which was used throughout this book, is given in Annex V.

NIAM models on the other hand lack the full rigour required to state complex object-role associations with all necessary, often logically demanding constraints. The EXPRESS modelling in STEP is much more precise in this direction. It is strongly object-oriented and hence does not require enumeration of semantically irreducible objects (entities) as primitives. Its use is easy for application modellers, even for structurally complex models. Different implementations of one model are readily derivable. But the integration of different application views in one model is more difficult in the object-oriented paradigm.

NEUTRABAS followed a convenient compromise by converting its NIAM models subsequently into EXPRESS models. They retained their basic semantically irreducible structure in this process while precise and complex constraints could be added as EXPRESS rules. Verbal definition of attributes was based on STEP style conventions. In the end the NEUTRABAS Partial Models were documented in EXPRESS and graphically in EXPRESS-G. The syntax of EXPRESS and the style rules for EXPRESS-G diagrams are described in the EXPRESS Language Reference Manual [3].

## 3.2 Ship Principal Characteristics Model

The Ship Principal Characteristics Model (SPCM) was developed in 1992 in Workpackage 1, Deliverable 1.2.6 [6] of NEUTRABAS. The status of the model must still be considered as a draft serving to illustrate the orientation and structure of this model. Further discussion is needed to prepare acceptance by a wider shipbuilding community beyond NEUTRABAS.

We define as principal characteristics of ships those properties which are suitable to characterize the ship and its functional capabilities from an overall viewpoint. Their relevance is not limited to any particular application area or life cycle stage. They can be pertinent as general information to any application context. In shipbuilding practice shipyards often prepare ship principal data sheets to document these principal characteristics for convenient reference. Practice varies, of course, as to what will be included in such documents. In the present version for a Ship Principal Characteristics Model it was attempted to include a broad variety of such information without trying to be exhaustive.

Engineering applications which may read or write principal characteristics information include the following:

- Hull form definition
- Hydrostatics, stability, trim
- Weights and centroids
- Freeboard
- Tonnage measurement
- Propulsion powering

- Hull and propeller performance
- Cargo hold and tank capacities
- Container capacities
- Loading and unloading
- Longitudinal strength

The following pages describe the SPCM by a set of EXPRESS-G diagrams. Annex VII.1 contains the EXPRESS representation of this model.

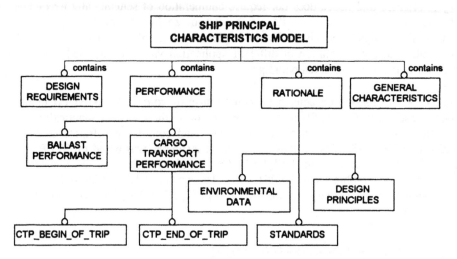

**Fig. 11**   EXPRESS-G Diagram of the Ship Principal Characteristics Model

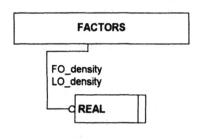

**Fig. 12**   EXPRESS-G Diagram of Factors

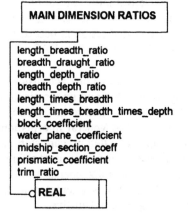

**Fig. 13**   EXPRESS-G Diagram of Main Dimension Ratios

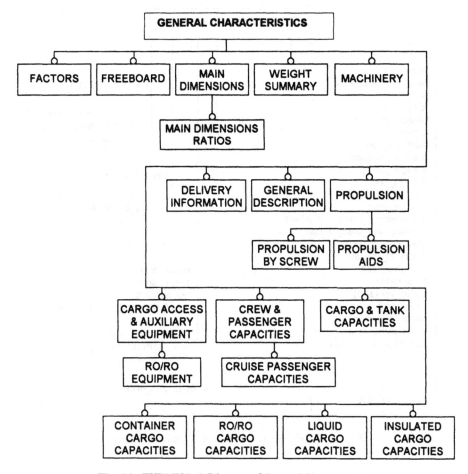

**Fig. 14** EXPRESS-G Diagram of General Characteristics

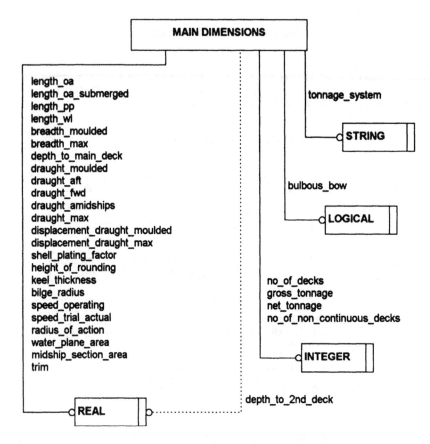

**Fig. 15** EXPRESS-G Diagram of Main Dimensions

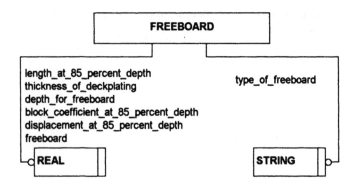

**Fig. 16** EXPRESS-G Diagram of Freeboard

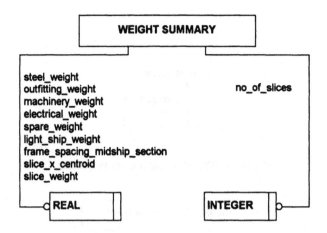

**Fig. 17** EXPRESS-G Diagram of Weight Summary

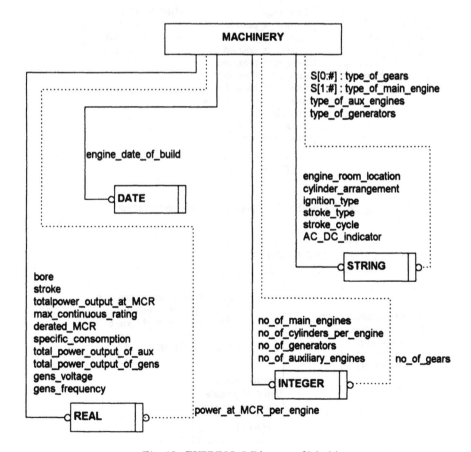

**Fig. 18** EXPRESS-G Diagram of Machinery

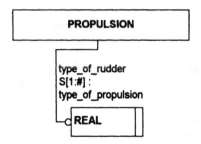

**Fig. 19** EXPRESS-G Diagram of Propulsion

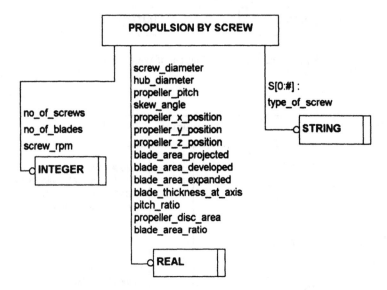

**Fig. 20** EXPRESS-G Diagram of Propulsion by Screw

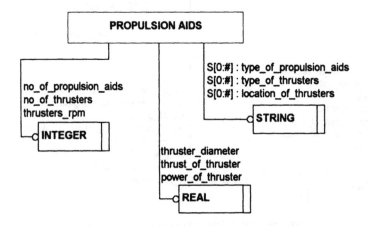

**Fig. 21** EXPRESS-G Diagram of Propulsion Aids

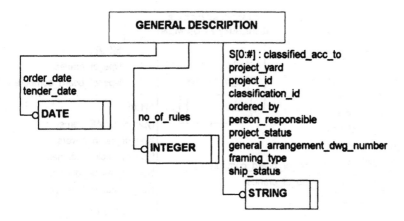

Fig. 22   EXPRESS-G Diagram of General Description

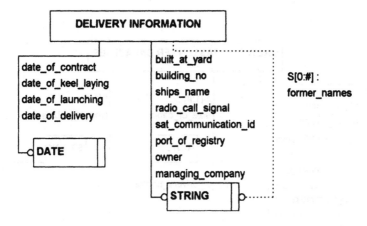

Fig. 23   EXPRESS-G Diagram of Delivery Information

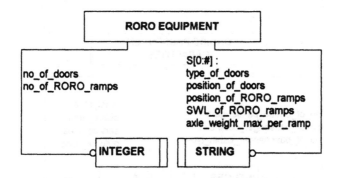

Fig. 24   EXPRESS-G Diagram of RoRo Equipment

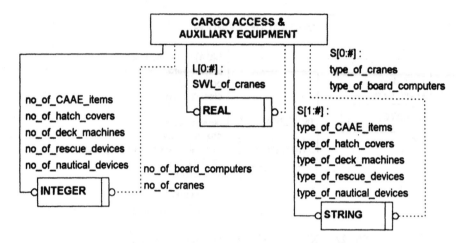

**Fig. 25** EXPRESS-G Diagram of Cargo Access & Auxiliary Equipment

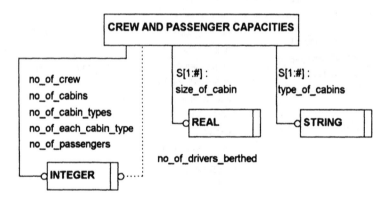

**Fig. 26** EXPRESS-G Diagram of Crew and Passenger Capacities

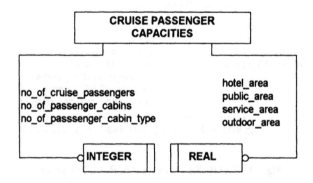

**Fig. 27** EXPRESS-G Diagram of Cruise Passenger Capacities

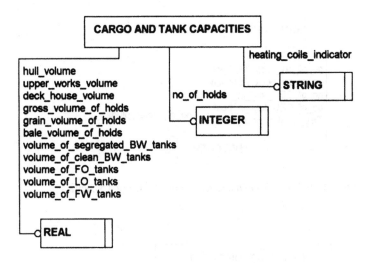

Fig. 28   EXPRESS-G Diagram of Cargo and Tank Capacities

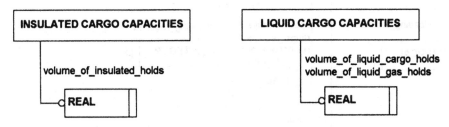

Fig. 29   EXPRESS-G Diagram of
          Insulated Cargo Capacities

Fig. 30   EXPRESS-G Diagram of
          Liquid Cargo Capacities

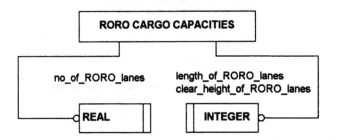

Fig. 31   EXPRESS-G Diagram of RoRo Cargo Capacities

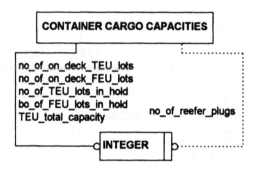

**Fig. 32** EXPRESS-G Diagram of Container Cargo Capacities

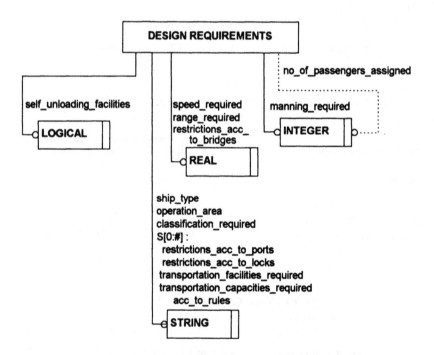

**Fig. 33** EXPRESS-G Diagram of Design Requirements

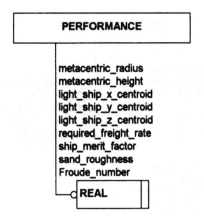

**Fig. 34**  EXPRESS-G Diagram of Performance

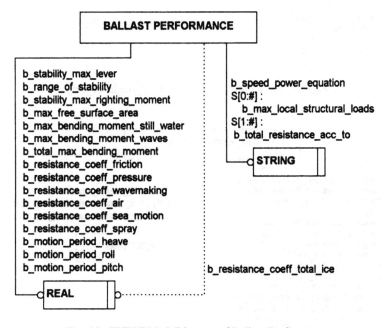

**Fig. 35**  EXPRESS-G Diagram of Ballast Performance

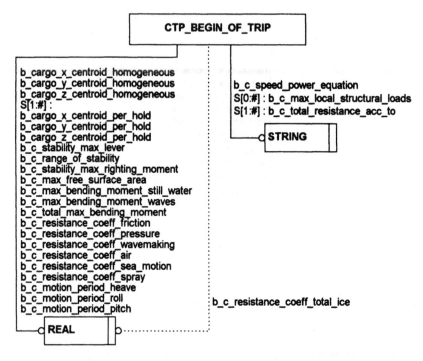

**Fig. 36** EXPRESS-G Diagram of CTP_Begin_of_Trip

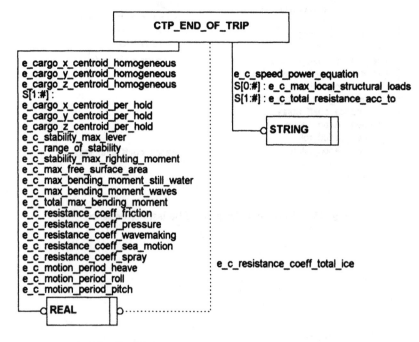

**Fig. 37** EXPRESS-G Diagram of CTP_End_of_Trip

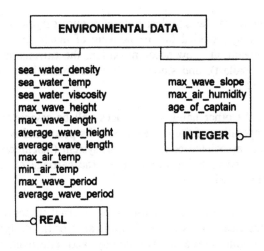

**Fig. 38** EXPRESS-G Diagram of Environmental Data

## 3.3 Spatial Arrangements

### 3.3.1 Introduction

The primary requirement of any marine vehicle is the ability to adequately perform some specified function or functions in the most efficient manner possible, with the efficiency being expressed in terms of economic performance, profit maximisation, operational safety and overall functionality. As can be appreciated, the often conflicting nature of the above desired features will usually lead to the development of a compromise solution which best suits the specified requirements for the design proposal [7].

In the context of the development of design proposals for marine vehicles in general, and displacement vessels in particular, it is considered that the development of the general layout arrangement is perhaps the most important of all of the activities associated with the design process. The design of the general arrangement is central to the complete design process with decisions taken at this stage having a far-reaching effect on subsequent stages of the design development process. For example, the overall internal layout of a vessel has a significant impact on the quality of the proposal in terms if its strength (local and global), stability, operational characteristics, economic performance and ease of production. The disposition of the various large structural items such as deck, transverse bulkheads, longitudinal bulkheads, flats, platforms, etc., therefore influences the ability of the vessel to perform the required functions in the most efficient manner possible.

As a result of their intersection with other surfaces, these large structural items combine to form enclosed spaces within the general confines of the overall hull envelope. These enclosed spaces (compartments) on board a vessel are used for various purposes depending upon the overall function of the vessel under consideration. For example, in the simplest case of a pure cargo-carrying vessel, the majority of spaces being provided for the housing of machinery items (main and auxiliary), for ship operation, crew accommodation, the carriage of fuel oil, diesel oil, water (fresh and ballast), and stores, with some spaces being left deliberately empty (cofferdams, etc.).

In naval vessels, on the other hand, most spaces will be associated with activities concerned with the operation of the vessel in both its peace and wartime roles, and for the housing of the associated equipment and system (machinery, domestic and weapons), with some spaces being provided for the carriage of the consumables required to sustain the vessel.

Although quite different in their specific use of the internal spaces available, both of the above vessel types clearly illustrate the significance of the design of the general arrangement in determining the overall quality and effectiveness of a design proposal, and hence the central position the general arrangement design activity has in the complete ship design process. In particular, the association of spaces, or compartments with functionality of the vessel is a key characteristic of this class of product, and distinguishes it from many other products such as mechanical, process and electronics. While compartments are created by the steel structure acting as boundaries, they also have the important property of containing systems, referred to as the outfitting. The outfitting systems may simply "pass through" a compartment, penetrating some of the boundaries, or may act as a service to the compartment as in the case of ventilation, power etc.

The significance of "space" in the design, construction, and operation of marine vehicles has meant that the NEUTRABAS project had to give special consideration to the modelling of space and its integration with both structure and outfitting. In practice the specification of the steel structure provided the main platform for the neutral specification of space; integration with the specification of outfitting systems was left to a subsequent stage.

### 3.3.2 Application Requirements

The requirements of the marine systems design application and the approach to modelling spatial arrangements was determined at an early stage of the project through a detailed review of existing modelling and design expertise and an analysis of a specific case study (mock-up). From this work the following requirements emerged:

### 3.3.2.1 Compartment Representation

The definition and description of internal enclosed spaces (compartments) is achieved through the representation of the relevant bounding surfaces. The relation of the surface definitions to the description of the compartments simplifies the process of creating compartment definitions and allows a topological description of the compartment to be adopted. The topological description however, refers to the topology of the boundary surfaces. There is also a requirement to try to have a direct representation of spatial topology such that adjacency relationships are maintained. Such modelling would support early design where exact geometries are poorly defined, design assessment where limitations on topologies are checked, and system routing where outfitting systems are routed through compartments. In practice most existing CAD systems represent the boundary information rather than spatial topology. Consequently a specification based on boundary geometry would provide the best common foundation.

In terms of the attributes which a particular compartment may possess, the following list was developed:

Entity : compartment

Attributes

| | |
|---|---|
| Compartment name | Required bulkhead tightness |
| Compartment number | Insulation category |
| Zone identifier | Noise category |
| Functional identifier | Illumination requirement |
| Actual deck area | Safety category |
| Minimum enclosed volume | Number of persons in compartment |
| Actual enclosed volume | Type of liquid in compartment |
| Minimum x dimension | Density of liquid in compartment |
| Actual x dimension | Requirements of liquid stowage |
| Minimum y dimension | (temperature, etc.) |
| Actual y dimension | Type of cargo carried in compartment |
| Minimum clear deck height | Stowage rate of cargo in compartment |
| Actual clear deck height | Requirement of cargo stowage |
| Maximum permissible acceleration | (temperature, etc.) |
| Air circulation rate | x centroid of compartment |
| Forward location limit | y centroid of compartment |
| Aft location limit | z centroid of compartment |
| Port location limit | Type of coating (tank) |
| Starboard location limit | Free surface moment |
| Lower location limit | Number of bounding surfaces |
| Upper location limit | Bounding surface identifiers |
| Access requirement | Adjacency requirements |

The above list clearly includes some attributes which cannot be possessed by all compartments. For example a compartment used solely for the carriage of liquid will not possess the attribute "number of persons in compartment". The above list is, however, intended to indicate all of the attributes which could be possessed by any of the compartments to be found on any type of vessel, including naval craft.

### 3.3.2.2 Space Representation

As a corollary to the definition of a compartment as a single volume bounded by surfaces, it can be stated that the total space of a ship is the sum of all the individual
compartment spaces. In practice, during design, it is often valuable to deal with
spaces which are made up of groups of compartments (indeed a ship is an example of
this), or are defined by physical or virtual boundaries. The need for the former is
reflected in the process of design which normally decides major subdivisions of the
bounding volume representing major functional divisions, before minor subdivisions.
Thus for example, accommodation space may be treated as a single entity initially,
until a breakdown into cabins and recreation areas is established. Even at later
design stages, this hierarchy of spaces is valuable as a way of grouping and
classifying space.

The need to use virtual boundaries is less obvious. However, these can be used in
design concept exploration in evaluating alternatives, and in production breakdown
structures where a physical break in the definition of a sub-assembly, block, or zone
would specifically avoid coincidences with major surfaces.

For all of the above reasons it was decided that there was a need to identify a product
concept of a "space" which is distinct from a "compartment". A compartment is itself
a space, but a space may contain a number of whole or partial compartments in its
definition.

Each compartment possesses individual functions. A space may thus be the accumulation of multitudinous functions.

### 3.3.2.3 Surface Representation

The discussion on the role of spaces already suggests a key requirement for surface
representations; a distinction between major and minor surfaces. This is largely an
application concept since both representations of surfaces will use the same resource
definitions from STEP. However, the application concept is valuable because of the
prime relationship of major surfaces firstly to the main functional subdivisions and
secondly to the major strength requirements of the structure.

Where a surface corresponds to a physical boundary, then that boundary will in
reality be made up of a complex of metal plates and stiffeners. This complex is of
significance to structural design and production; it is of less significance to spatial
design. To a large extent the spatial description can be satisfied with parametric
information about the surface. The key decision taken within NEUTRABAS was that
the prime definition of surfaces was through a mathematical representation already
available within the STEP resource models, and, where this surface corresponded to
a physical boundary, it would define the moulded surface of this boundary. The
concept of a moulded surface is one which is well understood by ship designers and
corresponds closely to existing working methods.

### 3.3.2.4 Reference System Representation

A system for positional reference emerged as a crucial requirement for the overall geometric description of marine vehicles. Existing STEP definitions included a limited definition of coordinate axes with reference to other coordinate systems. Some extensions to this were required for the ship application. In marine systems design and construction , a number of different systems are used simultaneously, and there is not a single uniform convention. An important requirement, therefore, in any data exchange, is to be able to transfer the definition of the default coordinate systems from one CAD system to another.

In general, coordinate systems in use are:

• Global Coordinate System which identifies a datum position and orthogonal coordinate axis system for locating the origin of vessel. Commonly the datum is at the lowest position of the aft perpendicular, with a line forward along the keel being the x-axis, upwards, amidships as the z-axis, and artwartships in a left-handed coordinate system as the y-axis. However many variations exist, including origin point amidships, y- and z-axis inverted, and using a right-handed coordinate system.

While a distance metric in real world coordinates is normally used, a frame count reference in the x-coordinate is also common.

• Local Coordinate System which applies to an assembly block or part or compartment or area covered by a drawing. Normally this would use the samecoordinates as the global coordinate system, but with a change of origin. However, for construction information an orientation of axis may also be used. In complex multihull offshore structures, a number of major local coordinate systems may be used to define the overall geometry, each one acting as a global system for its part of the design. The same approach would also apply to multihull ships.

### 3.3.3 Hull Spatial Organisational Model

The purpose of the NEUTRABAS Hull Spatial Organisation Model (HSOM) within NEUTRABAS is the definition of the geometrical and topological characteristics of the internal organisation of a vessel. This includes the definition of the internal enclosed spaces or compartments, the definition of the surfaces by which these internal spaces are bounded, the functions of the shipboard compartments, and the reference systems which serve to define the location and orientation of components of the product in 3-D space.

The model deals only with moulded geometry, which is the idealised geometry of all reference surfaces and curves of the ship without any structural members like plates and stiffeners being attached.

Figure 39 presents an overview NIAM diagram of the HSOM. It shows the basic structure with the main components of the model, the entities space, moulded-surfaces, and referential-system. This view emphasises the predominance of thespace or volume viewpoint in this model with the bounding surfaces attached to the spaces. The following figures are refined views of the principal entities.

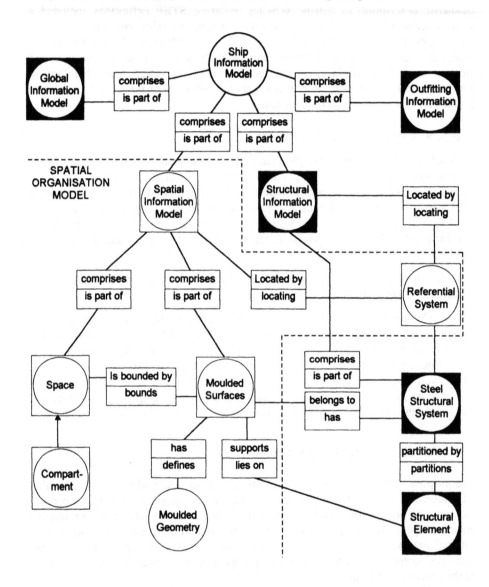

**Fig. 39** NIAM Diagram of Hull Spatial Organization

Figure 40 takes a closer view at moulded-surface. It shows the specific attributes which identify each surface. It also indicates how the description of moulded-surface can be mapped onto STEP generic resources in geometry.

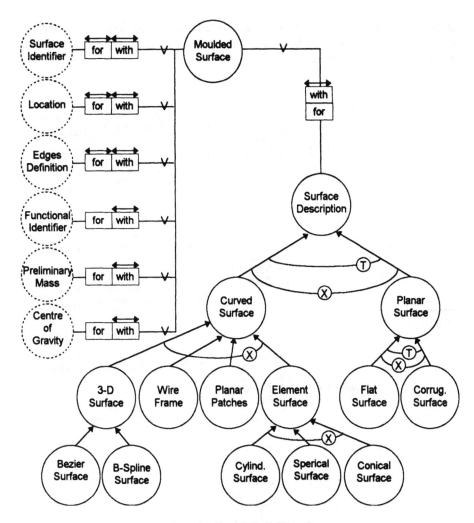

**Fig. 40**  NIAM Diagram of Moulded Surface Concept

Figure 41 explains the concept for the global reference system. This can be tied in with the methods by which geometry is founded in Part 43 of STEP.

The space and compartment concept is summarised in Figure 42. The entity space is carrying the information about the topological role of the volume in the internal organisation of the ship. The entity compartment which is directly associated with space is defined by those attributes which characterise the use and functional capacities of the volume. It is therefore mainly the space entity that resorts to generic resources of STEP geometry and topology.

The schema is thus organised in a well-structured way with only a few major entities which can serve as integration points while several minor ones do not complicate the pattern. The relationship between the major entities of this HSOM schema and the resources as well as the Ship Structural System Model, which shares some of the same resources, are clearly evident.

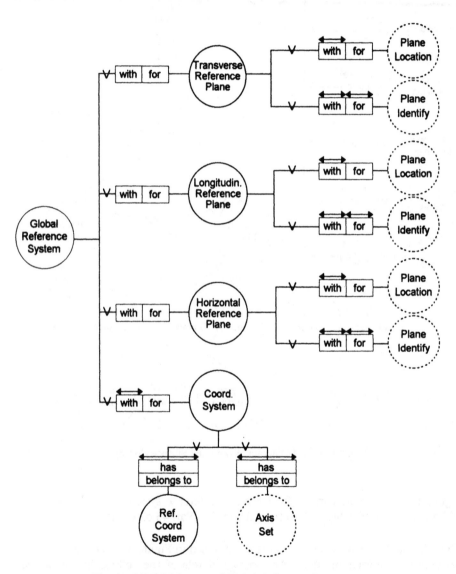

**Fig. 41**  NIAM Diagram of Global Reference System Concept

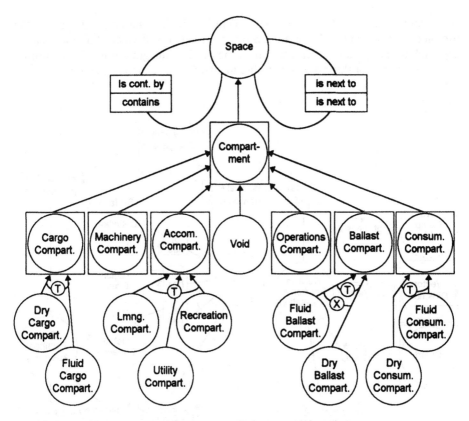

**Fig. 42** NIAM Diagram of Compartment Concept

### 3.3.4 Review

The definition of requirements and specification of the HSOM was developed largely as part of the steel definition work package of NEUTRABAS. As a consequence there is good integration achieved with the steelwork model. In the NIAM and EXPRESS definitions good accordance with the requirements was achieved. In many ways the specification exceeded the capabilities of many CAD systems in its use of space and topology. In particular the test cases were of limited scope and did not evaluate the topological aspects of the data exchange. Nonetheless considerable progress was achieved in developing this specification as a future model.

The rationale for developing a spatial model derives from the outfitting system requirements as much as from the steel work system requirements. In practice integration across these two models was not achieved. Figure 43 shows a NIAM model taken from the outfitting work package to describe the topological model developed. Reference within this to zones and regions can be found with relationships to major surfaces. This emphasis on zones and regions reflects a con-struction viewpoint of space as opposed to the design viewpoint taken in the structural model. Nonetheless, this point identifies the integration point between the two models. The work on Integration Strategy (Chapter 4) developed a more generic approach to integration which recognises the need to integrate through application resource schemas of ship volume and ship moulded surface amongst others.

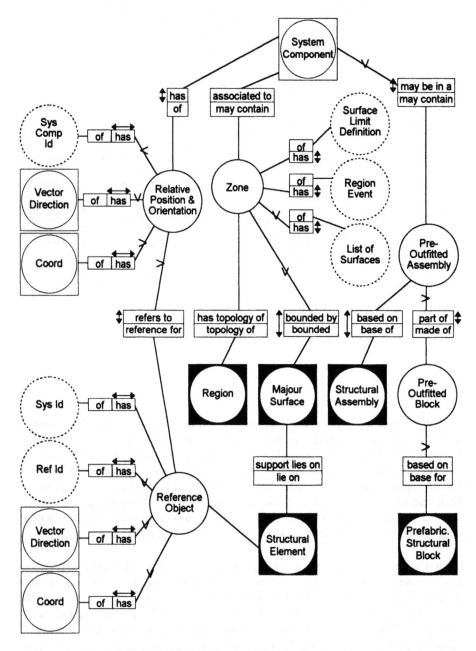

**Fig. 43** NIAM Diagram of System Component

## 3.4 Ship Structural Systems

### 3.4.1 Structural Information for Shipbuilding

#### 3.4.1.1 Ship Structures

The physical structure of a ship is a complex composition of elements that are very diverse in shapes and sizes, joined toqether by welding or other mechanical means.

Ship structures are a very good example of the type of structures that complex and multifunctional engineering products addressed by NEUTRABAS can have. A manufacturing plant, an aircraft or a complex piece of equipment usually have structures that are less varied, less complex or require less engineering attention than a ship hull structure.

Ship structures can be made of steel, light alloys, FRP or wood, or may be composed of two or more of these and other engineering materials. Structural components used in shipbuilding are generally formed plates and sections that can be rolled or built-up.

The ship structure constitutes the support and physical referential for all other ship sub-systems including the whole-ship-system, cargoes, machinery, accommodation, navigation and communications. In this respect, the ship structure is adapted to the needs of several other ship components, as well as the cargo and the crew.

#### 3.4.1.2 Meaning of the Ship Structure in the Ship Life-Cycle

Ship structures have some characteristics that make them live, complex engineering entities as compared to other complex structures that are created and defined with almost full independence of the rest of sub-systems of the total product. This live dependency is a major concern for shipbuilding and is present in the design and construction activities that are addressed by NEUTRABAS.

The life of the ship structure starts and ends with the ship herself. Therefore, the ship structure is repeatedly and progressively defined and changed along the ship's life as it is required by the ship's life-cycle. In particular, the ship structure lives along all the design stages, from concept to detailed engineering designs, and is manufactured and assembled, operated, maintained, repaired, modified and finally scrapped in permanent close relation with the whole ship and her sub-systems. During that life time functions performed by the structure change also.

The structure of a ship begins as a global concept with properties such as weight and centers, materials and main dimensions and shape. As the ship and her sub-systems get more and better defined the structure gains more attributes. In this process, the structure is treated as a referential to other ship sub-systems, and at the same time it is defined in its minor details to comply with the needs that those other sub-systems require from it.

### 3.4.1.3 Object Information Related to its Structure

At any operational stage the ship structure must be treated as a fixed, complete, complex entity which is very closely related to the object ship.

From the conception of a ship as an engineering product or object several global characteristics of it determine the requirements for its structure and at the same time they are determined by the solution of such structure.

Subdivision, spatial arrangement, foundations, rigid surfaces, weight, center of gravity, inertia, static and dynamic global behaviour and responses of the ship are governed by her structure.

Systems within the ship, equipment, machinery, outfitting and all sorts of cargo and other elements on board have to be supported by, or attached to, or be referred or related to some element or part of the ship structure.

### 3.4.1.4 Applications Related to Ship Structures

NEUTRABAS is to be used to support all conceivable sorts of uses and applications that need some data pertaining to any item of the hull structure of a ship or equivalent complex product. Therefore, such structure must be complete, i.e. it must consist of all parts that belong to the hull of the ship.

On the other hand, such parts must be accessible in all the ways needed by all possible applications using the product, i.e., they shall be given all the attributes necessary for every different application.

Conceivable applications (A) along the product life are listed in Annex II. All of them may make use of any structure component either singly or combined with others.

In order to present a chronologically sequential list of applications, the life-cycle of the product is analyzed by breaking it down to accepted stages or periods which are common to shipbuilding.

In some cases, attributes or predicates (P) are listed besides the applications in order to complete a better description of every stage.

In other cases, groups of applications are included in one single concept to be taken as a category (C).

### 3.4.1.5  Relations to Other Ship Sub-Systems

Ship structures are related to other ship sub-systems by the application of functions such as support, attachment, foundation, reference, loads and other actions that can be imposed on the structure by those functions.

Relationships of the ship structure to other ship subsystems are defined within the scope of the whole Ship Information Model.

### 3.4.2  Structural Information Model Requirements

The product high level model provides the framework for product data modeling. The product model has the meaning of a concept that can be defined, represented and also presented in different ways, as requested by the user. Any concept can be defined in terms of its statistics, its functions and its changes. Changes in the concept are its statistics along its life-cycle. In ships, changes refer to life stages such as design, engineering and production.

The same product model changes along its life stages and is the object of operations performed on it by different activities such as design, manufacturing, management and planning.

A ship as a product or system is composed of many interrelated sub-systems and elements. She can also be defined in terms of functions and relationships as an information model. Both the system model and the information model are described or specified using smaller, elementary pieces of information with physical, functional and operational meaning associated with them. These elementary components are the modeling resources that are specified as resource models.

### 3.4.2.1  User Representation of the Structure

A main goal of NEUTRABAS is to provide support for any user representation of the ship structure that may be required by application software systems. Since user representations may be graphical, literal or electronic, the structure has to be modeled by NEUTRABAS with all possible applications in mind, detailed and interpretable by current and future data processors.

User representations of the ship structure are necessarily related to user representations of the ship as a complete product, and to operational characteristics of the ship that are a function of her spatial organization.

Different user applications on the ship structure will require different views of the structure. Optimization of such applications will be better achieved using particular representations of the ship structure. The objective of NEUTRABAS is to define a single, neutral, complete model for the structure; therefore a decision has to be made on a certain representation of the structure that includes a certain relationship between its components. This model has to represent both physical and functional dependencies and has to be applicable to any user application.

Defining such a model has many solutions. The best solution can only be assessed at the application stage. NEUTRABAS has chosen a physical, top-down representation of the ship structure which enables the user to check its degree of definition throughout the design process. It is included and explained in Subsection 3.4.4.

This Steel Structure Representation Diagram represents and reflects a definition view of the problem based on topological levels and physical relationships pertaining to Ship Design. This approach should be valid for checking the completion of the ship structure definition as a physical reality, but without any relation to functionalities or, in the sense presented in 3.4.1, without a direct provision for categorization of attributes with regard to foreseen and not-foreseen applications. Moreover, this approach facilitates the declaration of EXPRESS statements referring to connectivity and physical interdependency.

This notwithstanding the accepted approach should not be taken as a constraint on the Data Base Structure in the sense that it prejudices a certain type of hierarchical relationship among Information Model Components.

### 3.4.2.2 Physical Entities

The NEUTRABAS Structural Information Model was developed using a representative part of the ship structure. For an overview see also the list on top of p. 47.

It was described by means of NIAM diagrams first. A top-down modeling approach was followed. This partial model of the structure is used as an illustration of solutions proposed by NEUTRABAS. Whenever necessary, this model was checked, complemented or modified applying NIAM to a bottom-up approach. The corresponding set of entities was checked for compliance with STEP and some new entities were added when the proposed STEP schema was found deficient.

### 3.4.2.3 Geometry and Topology

Geometry and Topology are fundamental characteristics of the NEUTRABAS Ship Information Model. As it is expressed clearly in the NIAM Diagram in Fig. 39 the Spatial Organization Model comprises all entities needed for reference in the Structure Information Model. Spaces, Moulded Surfaces and Referential Systems are used by the Steel Structural System and all Structural Elements that are supported in it.

The structure of the ship hull has several attributes which are in common to the hull from a global perception of the ship and which are liable to be used for applications other than design and construction of the structure. These attributes include major surfaces, referential system and bounded spaces among others.

* Steel Structural System Entities sample relationships:

Major Surface                              Stiffening Element
Structural Element                         Stiffening Element
Primary Substructure                                          Assembly
Parts Assembly                               Composite Stiffener
  Parts Sub-Assembly                         Commercial Stiffener
Plate Sheet                                  Transition Part
  Joint                                      String
  String                                     Standard Library
Plate Assembly                             Commercial Stiffener
  Joint                                      Standard Library
Plate                                        Endcut
  Joint                                      Connecting Part
Prefabricated Block                          Structural Opening
  Prefabricated Sub-Block                    Joint
Prefabricated Sub-Block                    Composite Stiffener
Large Opening                                Standard Library
  Standard                                   Endcut
Structural Opening                           Connecting Part
  Standard                                   Structural Opening
  Access/Lightening Hole                     Joint
  System Penetration Hole                Connecting Part
  Distribution System Part             Endcut
  Cutout Hole                              Parametric Endcut
  Structural Penetration                   Non-Parametric Endcut
    Hole                                     Standard Library
  Air Escape                             Transition Part
  Fluid Opening
  Rathole

From the point of view of the designer it would be more convenient to use a single, unique definition of those attributes throughout the design process. Therefore, those attributes can be made part of some sort of entities from which they would be borrowed, not only by the structure but by other entities related to other views of the ship as a live product.

It seems convenient for the goals of NEUTRABAS to separate categories of attributes such as Geometry and Topology from their use by the structure, the outfitting or the ship designer/naval architect in order not to multiply the number of attributes and their definition, and at the same time to help keeping coherence within NEUTRABAS.

Reference systems can be nested at different levels and be used for different applications, e g. the definition or derivation of structural entities. Business rules associated with each one can describe the application of the system and how it is defined. They can also be used with limited ranges of validity for any system, especially for local ones to a certain area, space or sub-structure. Their attributes will thus permit checking the coherence and soundness of the specification of unit entities to be derived or defined using those reference systems.

However, when describing a certain axis for a local coordinate reference system, one may use other entities to derive such a line. A sample of this will be the case of a transverse bulkhead, where a certain physical line is chosen to be used as a base line for locating some structural entities such as stringers, stiffeners and other elements or connections. That physical line may be an edge, or an intersection of two planar entities.

In cases like this it is not practical calculating the unit vectors of the resulting axes in a global reference system, since applications using the transverse bulkhead are going to refer to those local axes, or they would have not taken the trouble to define them.

From these considerations, it follows that a reference system specification should consist of sets of attributes whose application and meaning are to be governed by business rules that may prescribe the complementarity or the exclusiveness of such sets.

Major surfaces are defined with their own local system of reference. This system should be stated with reference to some other system, be it the global one or simply a parent system.

Wire frame models are used for a very wide range of applications and many CAD/CAM systems today use such simplified definition of the ship hull. Certain design applications for naval architecture calculations and drafting rely on a wire model rather than on a continuous 3D surface. This preference is sometimes justified by the use of small desk-top computers and screens when simple, quick processing has priority over accuracy. The use of wire frame models requires defining the type of curves to be supported.

A valid set of such curves may be: planar 2D lines; spatial 3D lines; intersecting surfaces. Curved surfaces should include as a sub-type POLYHEDRAL surface or any other type of PATCHED approximation that may be used for different applications, e.g. planar elements for FEM, hydrodynamics, integration or expansions for fabrication.

As an example of the role of the entities that are used by the Structure Information System, the complete treatment and specification for Moulded Surface is detailed here. It consists of a threefold definition of its attributes and meaning: using natural language sentences, translated into a NIAM diagram and coded in EXPRESS.

```
* Entity Name : MOULDED SURFACE
```

A MOULDED SURFACE is a geometric boundary within the overall envelope of the product. A MOULDED SURFACE may provide the topological and geometrical references for an associated structural boundary. A MOULDED SURFACE may form part of a boundary of a physical or hypothetical enclosed space within the product spatial topology.

Natural Language:

A MOULDED SURFACE

> is part of a SPATIAL INFORMATION MODEL
> may form part of the boundary of a SPACE
> may support a STRUCTURAL ELEMENT
> has MOULDED GEOMETRY
> has a unique SURFACE IDENTIFIER
> has a unique LOCATION
> has one FUNCTIONAL IDENTIFIER
> has zero or one PRELIMINARY MASS
> has zero or one CENTRE OF GRAVITY
> has one SURFACE DESCRIPTION
> has one unique EDGES DEFINITION

NIAM Specification is shown in Fig. 40

EXPRESS Specification:

```
ENTITY moulded_surface;
    has_surface_identifier:     surface_identifier;
    has_moulded_geometry:       moulded_geometry;
    has_location:               location;
    has_functional_identifier:  functional_identifier;
    has_preliminary_mass:       OPTIONAL preliminary_mass;
    has_centre_of_gravity:      OPTIONAL centre_of_gravity;
    has_surface_description:    surface_description;
    has_edges_definition:       edges_definition;
  UNIQUE
    has_surface_identifier;
    has_location;
END ENTITY
```

### 3.4.2.4 Boundaries and Referentials in Ship Design

Use of boundaries and referentials in the design of ships and ship structures follows some procedures that have been well established in the past by naval architects. These conventions demand from NEUTRABAS a special attention to the significance and roles of the Boundary and Referential concepts which must be supported by the solutions proposed by NEUTRABAS.

A brief discussion on the use of boundaries and references in ship design is included as Annex III to this Section to help the non-shipbuilding reader appreciate the value added by those two concepts on ship design.

### 3.4.2.5  Functional Characteristics

The NEUTRABAS approach to modeling of ship structures and their components requires that their models be used beyond the mere physical representation of the objects. This calls for a functional capability of the models as it can be found in ISO. The STEP approach does not fulfill all of the NEUTRABAS goals with respect to the ship structure, however it is a valid and thorough approach to the definition of an Object Oriented Data Structure, which is acceptable as a pattern to be followed by NEUTRABAS.

Although a set of "functionality requirements" has been defined for NEUTRABAS, such set shall be considered as not limited, open to the system programmer and criteria and rules shall be provided to facilitate the maintenance of a particular functionality at any given user installation. Therefore the list presented in Annex II should be taken as an open set.

### 3.4.2.6  Relation to Other Ship Partial Models

Since the NEUTRABAS Ship Information Model is divided into a number of sub-models which describe the ship throughout its life-cycle, the Structural System Information Model is closely related to the other models.

It uses the Spatial Organization System that provides the support and integrity in which the Outfitting System Model is defined. This integrating capability is an extension to previous STEP scopes and will provide concepts to represent other products of the STEP-AEC industry activity.

### 3.4.2.7  Shipbuilding Standards and Norms

Shipbuilding standards are solutions for parts and construction details for the structure that are defined or accepted by a shipyard. They determine the dimensions and scantlings of parts that are repeatedly used in the structure and the connections between two or more parts. They are expressed by a set of parameters and their relationships that can be used by design oriented applications. Standards are stored in libraries and instantiated as parts by means of values of parameters that define them.

Standards in shipbuilding take a different form and meaning in design and manufacturing. They are normally used to resolve connections, openings, attachments, reinforcements or connecting parts and details of these.

The simplest method of exchanging standard parts between two application systems is to treat them as external references. Their identification triggers a special-purpose code that resolves the relation between their parameter values. The exchange is then dependent on the code and is treated differently by each side of the transaction. This method of exchange requires that standards be stored as algorithms that encode the rules for their application and be able to be referred to by both the source and the target systems.

Other possible ways to transfer standard parts are: geometry transfer; table driven transfer and neutral library format (STEP). For a full implementation NEUTRABAS recommends a method using geometry combined with external references.

Most substructures of the ship hull follow accepted patterns that define both their structural arrangement and the detail solution taken for them. These characteristics of element arrangement and details of their connections, ends, and welds constitute a type of structure. Thus, for instance Transverse Bulkheads for certain ship types are arranged using a type of structure : flat or corrugated, number of stringers, stiffener ends.

The use of this type as an atribute to the structural entity transverse bulkhead will allow changes in the solution of a hull structural design without altering the rest of the attributes at the same level of information.

To support this idea, a Library of Standard Structures will have to be generated as higher level norms that could be invoked with attributes for the entities using them.

A geometrical model is the final step in the definition of a structural detail or norm. Any structural solution starts at a concept level and is followed by a mechanical model that includes physical solutions and engineering calculations as a means to arrive at good technical solutions. Every solution has some "why's" upstream in the design process.

In order to support the ship's life-cycle from concept definition on through disposal other higher level Standards/Norms/Rules need to be dealt with by NEUTRABAS, so that they:

• can define the existence of a structural element, its type and spatial position;
• can define the type of elements connection, to be modeled in the technical solutions and calculations; e.g. free or fixed ends;
• can refer to the calculation process and rules when necessary

Based on this approach, at least two types of standards can be defined at a higher level:

• concept standards, which define further modeling conceptually;
• calculation standards, which support modeling mathematically.

### 3.4.3 NEUTRABAS Approach as Related to STEP

Structure Information Models in STEP/NIDDESC and NEUTRABAS show differences that are the results of the differences in the two approaches. Small differences exist in some entities and attributes and at the level of details, but they stem from the type of applications that have been foreseen for NEUTRABAS.

The main differences between the two models can be better shown by comparing their scope of information and their global NIAM representation.

As Figs. 44, 45, 46 and 3 show, the scope of the NEUTRABAS Information Model is larger than that of STEP/NIDDESC. This means that some new entities, relationships and attributes have been created by NEUTRABAS that did not yet exist in STEP/NIDDESC. They are needed by some new applications that NEUTRABAS is to support, particularly in the areas of:

– preliminary design;
– functional design;
– production engineering;
– production planning;
– strength and stability, costs.

NIAM diagrams for STEP/NIDDESC and NEUTRABAS Models also show these differences (Figs. 47 and 48a/b). Although the two models have similar diagrams at the level of detail such as plates, stiffeners and structural openings, the NEUTRABAS diagram includes a new intermediate set of entities between the "Steel Structure System" level and the "Structural Part" level, which have been found necessary to support some applications, in particular:

preliminary and functional design;
production engineering;
production control and production
planning.

### 3.4.4  Specifications for Ship Hull Structure Modeling

#### 3.4.4.1  Hull Structure Representations: Top-Down/Bottom-Up

Ship structures are represented or described in different forms by different users. The structural designer uses a top-down approach. It starts by a global perception of the structure and proceeds to define functionally significant structural assemblies. Then these assemblies are defined in detail to their individual constituents or elements. Finally, constructional data and details are defined.

From the manufacturing point of view, the approach is rather from bottom up. Structural assemblies are built-up from smaller components and are then joined to other ones to finally constitute the complete ship.

Using a bottom-up approach to define the structure will be equivalent to defining individual components, each of them with all its attributes and functionalities. Later on, some integrating device should pick every one of these components and by relating its definition to others would formulate parent-brother-son relationships and detect the relating functionalities attached to their attributes. This approach would consist in the definition of a sort of bugs which, once put into the common bag

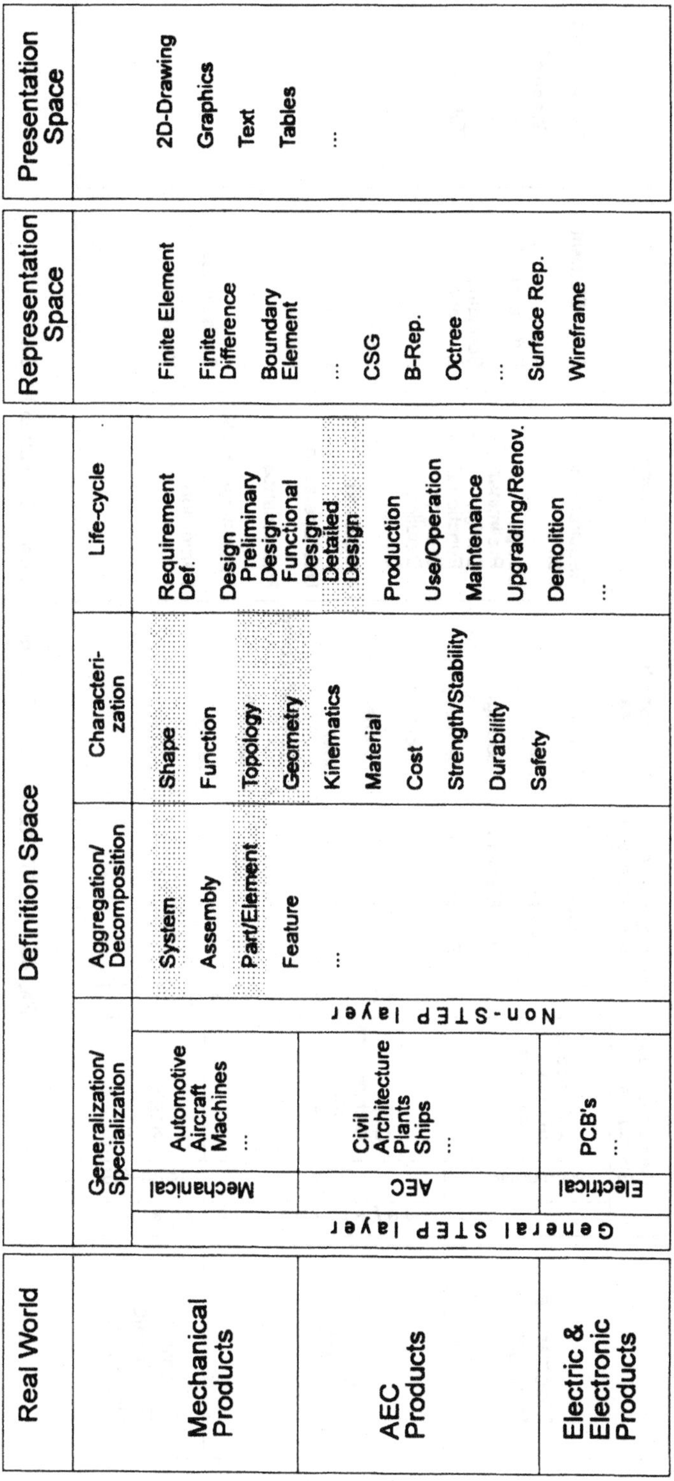

| Real World | Definition Space | | | | Representation Space | Presentation Space |
|---|---|---|---|---|---|---|
| | **General STEP layer** — Generalization/Specialization | **Non-STEP layer** — Aggregation/Decomposition | Characterization | Life-cycle | | |
| **Mechanical Products** | Mechanical: Automotive, Aircraft, Machines, ... | System | Shape | Requirement Def. | Finite Element | 2D-Drawing |
| | | Assembly | Function | Design Preliminary Design | Finite Difference | Graphics |
| **AEC Products** | AEC: Civil, Architecture, Plants, Ships, ... | Part/Element | Topology | Functional Design | Boundary Element | Text |
| | | Feature | Geometry | Detailed Design | ... | Tables |
| | | ... | Kinematics | Production | CSG | ... |
| **Electric & Electronic Products** | Electrical: PCB's, ... | | Material | Use/Operation | B-Rep. | |
| | | | Cost | Maintenance | Octree | |
| | | | Strength/Stability | Upgrading/Renov. | ... | |
| | | | Durability | Demolition | Surface Rep. | |
| | | | Safety | ... | Wireframe | |

Fig. 44 STEP/NIDDESC - Fluid Structural System Information Model

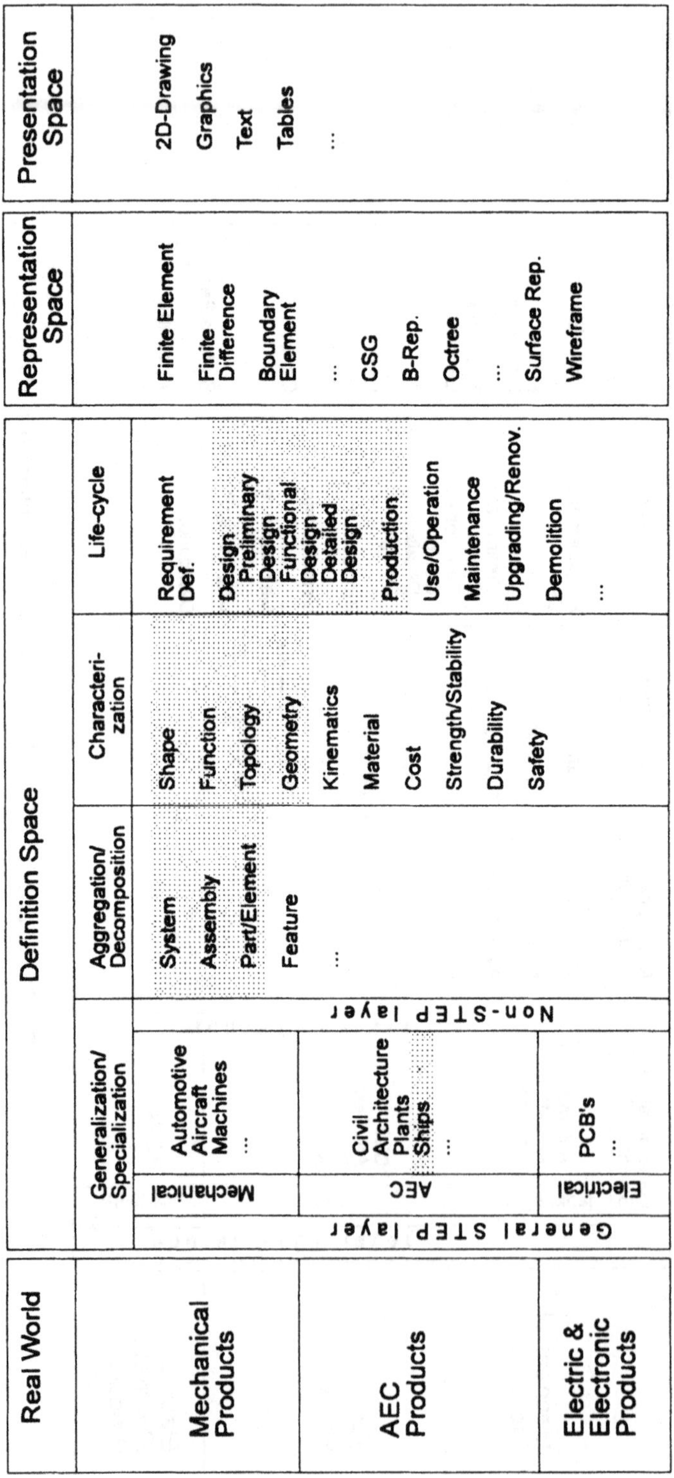

**Real World**

Mechanical Products

AEC Products

Electric & Electronic Products

**Definition Space**

*Generalization/Specialization*

General STEP layer

| Mechanical | AEC | Electrical |
|---|---|---|
| Automotive Aircraft Machines ... | Civil Architecture Plants Ships ... | PCB's ... |

Non-STEP layer

*Aggregation/Decomposition*

System
Assembly
Part/Element
Feature
...

*Characterization*

Shape
Function
Topology
Geometry
Kinematics
Material
Cost
Strength/Stability
Durability
Safety

*Life-cycle*

Requirement Def.
Design
  Preliminary Design
  Functional Design
  Detailed Design
Production
Use/Operation
Maintenance
Upgrading/Renov.
Demolition
...

**Representation Space**

Finite Element
Finite Difference
Boundary Element
...
CSG
B-Rep.
Octree
...
Surface Rep.
Wireframe

**Presentation Space**

2D-Drawing
Graphics
Text
Tables
...

**Fig. 45**   NEUTRABAS - Ship Structural System Information Model

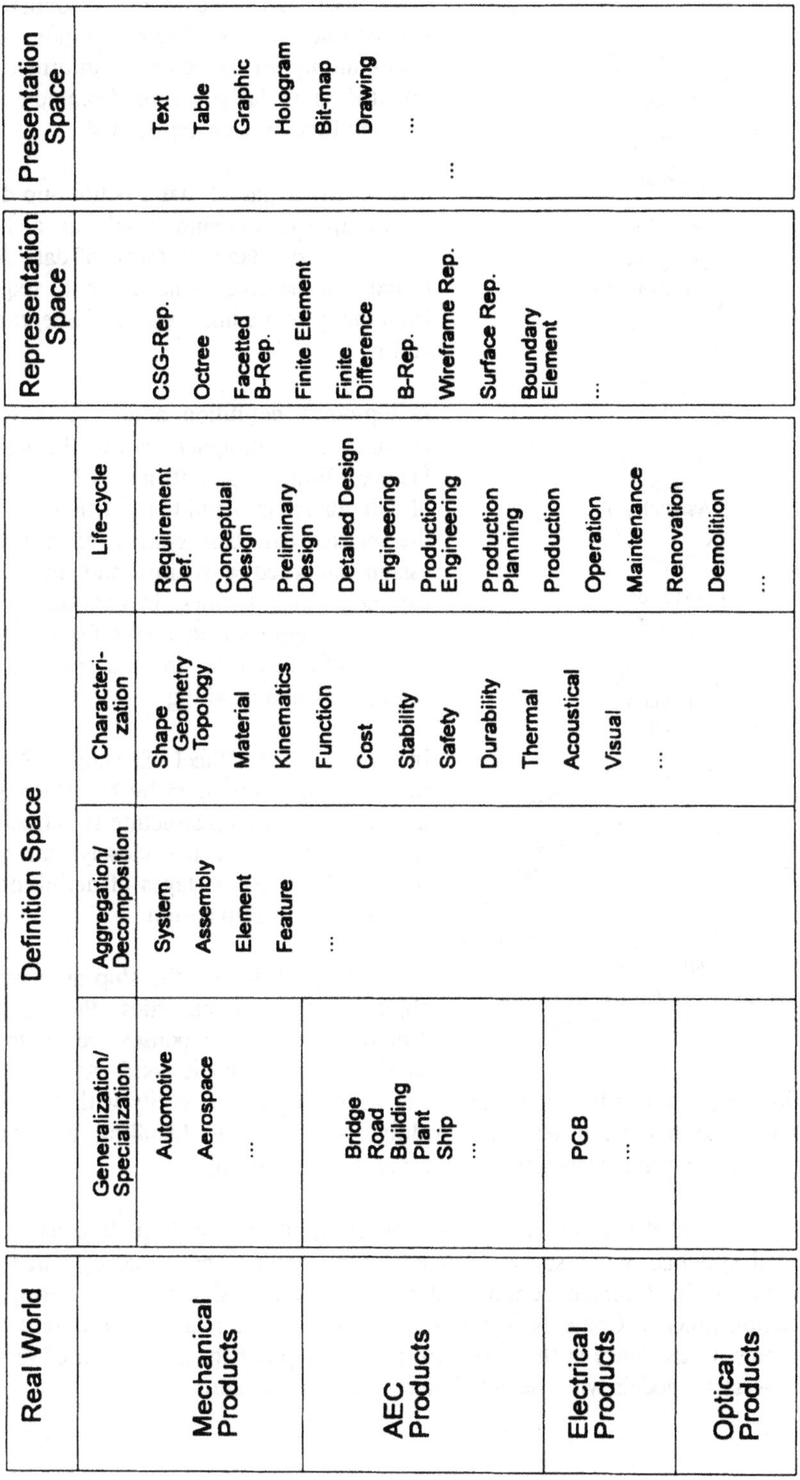

| Real World | Definition Space | | | | Representation Space | Presentation Space |
|---|---|---|---|---|---|---|
| | Generalization/ Specialization | Aggregation/ Decomposition | Characteri- zation | Life-cycle | | |
| Mechanical Products | Automotive Aerospace ... | System Assembly Element Feature ... | Shape Geometry Topology Material Kinematics Function Cost Stability Safety Durability Thermal Acoustical Visual ... | Requirement Def. Conceptual Design Preliminary Design Detailed Design Engineering Production Engineering Production Planning Production Operation Maintenance Renovation Demolition ... | CSG-Rep. Octree Facetted B-Rep. Finite Element Finite Difference B-Rep. Wireframe Rep. Surface Rep. Boundary Element ... | Text Table Graphic Hologram Bit-map Drawing ... |
| AEC Products | Bridge Road Building Plant Ship ... | | | | | |
| Electrical Products | PCB ... | | | | | |
| Optical Products | | | | | | |

**Fig. 46   STEP Planning Model**

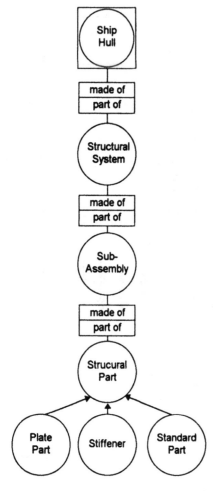

**Fig. 47**  STEP/NIDDESC
Simplified Global NIAM for
Ship Structural System

of NEUTRABAS, would lively connect to one another through their common legs, thus building up the complete information-oriented, physical structure. Common legs are identified by each application.

It is readily apparent that a bottom-up definition approach would result in a high degree of redundancy in terms of data definition. On the other hand, it would require from the system a high degree of integration ability.

A top-down definition approach is more familiar to the designer than to the manufacturer. While it permits a very low degree of redundancy in definition it demands from the selected data base system the ability to assign inherited attributes and to detect common functionalities. In contrast to the bottom-up approach this calls for the derivation of attributes in the process of segregating and detailing.

In Annex I a "Steel Structure Representation Diagram" used by NEUTRABAS is included. The ship structure is defined by levels of information that closely reproduce the actual phases or stages of definition of the structure in ship design.

The first level defines the ship as a whole object, with characteristics that permit reproducing her responses to external causes as a deformable solid. At this level the ship is represented by her behaviour as a multifunctional entity with attributes that belong to her main features, loadings and stability, handling and cargo conditioning, crew and accommodation, propulsion, outfitting, etc.

The second level deals with the spatial organization of the ship. It consists of: referential systems; major surfaces organization; volume and space organization. Information at this level can be modified during the ship design and even during her construction process. Contents of this information can be related to different ship sub-systems. Consistency at this level will provide support for integration of several ship sub-system models with the one representinq her structure.

The third level is for the ship structure. It includes the global perception of the structure that is related to the ship entity and to its spatial organization. Information at this level supports the definition of structural data that is derived from the two upper levels and is generally defined during the design stage. This could be considered as the result of integration of fourth level entities.

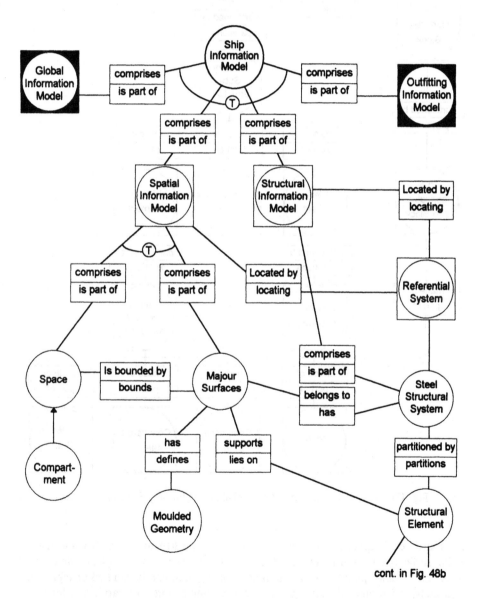

**Fig. 48a**   NEUTRABAS Simplified Global NIAM for Ship Structural System

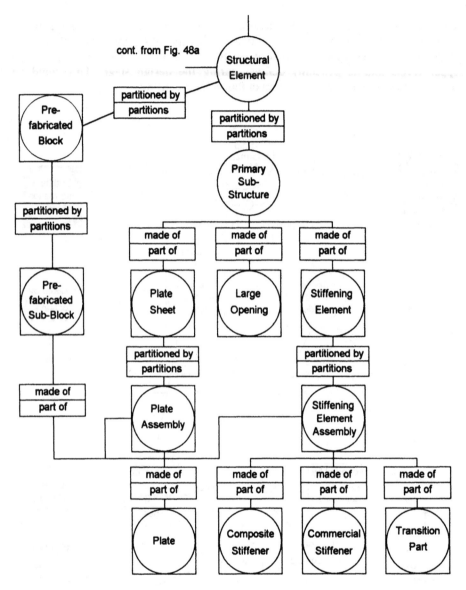

**Fig. 48b**  NEUTRABAS Simplified Global NIAM for Ship Structural System

The fourth level represents the general partition of the structure, with two sub-levels: major surfaces partition (4.1) and structural and spatial partition (4.2). Major surfaces define major joints and structural components that can be studied or fabricated separately. Partition of the structure into major components will vary from design to design and more so from yard to yard. Information at this level is spatially related through the second level.

General and detail structural representation is included at the fifth level, which consists of three sub-levels: shell plating (5.1), stiffening (5.2) and openings (5.3). Shell plating is related to major surfaces at the fourth level for the definition of assemblies, primary substructures and prefabricated blocks. Information about plating sheets includes their boundaries and joints. Stiffening elements include information about construction standards and connections. Apertures in structural elements include all sorts of large openings for circulation of persons, fluids or passage of any other solid elements.

Finally, a sixth level of representation includes all sorts of particular details that generally relate the structure to machinery and sub-systems within the ship. These are details or separate entities that can be attached to or detached from the rest of the structure as rather independent units. Design standards, bilge wells, and machinery foundations can be included at this level.

Although the diagram included in Annex I that depicts this organization is a representation of the design process it shows many connections between a great number of elements at different levels. These connections will grow in number and will change if another category of applications is represented, for instance the fabrication process for different yards. The possible extent of these other views can be derived from the review of NEUTRABAS applications included as an example in Annex II. It has been an early task of NEUTRABAS to ascertain that this information analysis can be represented in conformity with STEP and that it can be described using EXPRESS.

### 3.4.4.2 Entities and Attributes

The ship structural system is decomposed into entities of decreasing complexity and increasing generality. Attributes define physical and functional characteristics and are used to relate every entity to some application or view of the user. A list of entities used in NEUTRABAS is included in Annex IV.

### 3.4.4.3 Models and Architecture

NEUTRABAS structure information model has been developed using a methodology that combines NIAM, STEP and Object Oriented Methodologies. Knowledge acquisition and information structure have been described using natural language. NIAM has been used as a baseline modeling method but has not been applied to the detailed and complete model description. Wherever it has been found applicable the STEP\NIDDESC Application Reference Model for Ship Structural System has been referred to and used, but some information concerned with the topology and views of the object required by NEUTRABAS differ from the STEP Working Draft for Ship Structures [1].

The structural system is modeled by a high level model that includes the concept and integrates the information models for product, activities, design, production engineering, manufacturing and management.

### 3.4.4.4  Natural Language Descriptions

Natural language descriptions of objects are used to represent the information in a way that is understandable to users in the fields of ship design and shipbuilding. They are the result of communication between experts and modelers. This provides a first record of the analysis stage that can be formalized at a later stage. It includes detail information on the concepts of the objects, how to use them and how to operate or manipulate them. These are the starting points for NIAM and EXPRESS representations.

Objects are described with a set of static features as attributes, and a set of dynamic features or operations which define their behaviour.

Description rules in natural language call for a complete vocabulary of terms, sentences written with a limited syntax and just a static view of the object, since target representations in NIAM and EXPRESS do not have a dynamic capability.

Some natural language descriptions that are representative of the Ship Structural Information System are included in Annex VI.

### 3.4.4.5  NIAM Diagrams

The binary entity-attribute relationship model in [NIAM89] specifies binary relationships (facts) between objects. Objects are abstract or tangible entities that can be lexical (LOT) or no-lexical (NOLOT) and they are related by BRIDGES and IDEAS. Bridges between LOT-NOLOT define attributes of entities. Ideas between two NOLOT define relationships between two corresponding entities sets.

Some selected NIAM diagrams developed for the ship structural system in NEUTRABAS are included in Annex VI to illustrate the Information Models. They are accompanied by the natural language description of the same concepts. From the experience gained in this project it was concluded that certain tools would help much in adopting NIAM diagrams to other modeling projects. These tools could be: a graphic editor to enable handling subsets of the model; a parser to perform syntactic and semantic checks; a database manager to prove that NIAM models correspond to what is needed at instance levels.

### 3.4.4.6  EXPRESS Language Codes

Codes written in EXPRESS add to the NIAM diagrams some information that is needed to handle data to be stored in databases. However, a language like NIAM lacks the dynamic capability to model functions, behaviour and changes in the concept or entity.

The EXPRESS language is used in NEUTRABAS to:

– provide a different view of objects according to different classifications of data;
– provide different levels of abstraction (semantics)
– support the use of inheritance;
– provide granular information to facilitate changes.

The EXPRESS language codes for the same NIAM diagrams in Annex VI are included in Annex VII for reference.

### 3.4.4.7 Resource Models

In STEP architecture resource models are defined as independent models which can be used by other higher level models. Each resource model is concerned with a specific area of modeling. Geometry, topology, material, time, date are some of the resources used by NEUTRABAS from the series of STEP Integrated Generic Resource Parts.

## 3.5 Outfitting Systems

### 3.5.1 Overview

In NEUTRABAS the Ship Outfitting Systems Model (SOSM) is one of the four major partial models of the Ship Information Model (Fig. 1). Together with the Global, Spatial and Structural Information Models it forms a set that endeavours to comprise the full ship product information as completely as possible.

The outfitting systems of a ship form a very comprehensive set of shipboard functional systems with which the bare hull structure of the ship must be equipped in order to realize all operational missions of the vessel. It is evident that these systems together constitute a major share of the product value in a ship. Fig. 49 gives a basic overview of the type of outfitting systems that belong to this modelling domain. In NEUTRABAS Workpackage 4 was responsible for modeling the ship outfitting domain. Obviously it was not possible to model all outfitting systems in every detail, so priorities had to be placed to limit the model scope.

The primary emphasis in the NEUTRABAS outfitting product models was placed on the early design stage of these systems, which includes the life-cycle stages of requirements definition, contract design and functional design with lesser emphasis on detailed engineering design and production planning although some of these aspects may be relevant, too. This explains why geometry (size, location), topology (connectivity) and function (role) are included in the early design models of these systems, but e.g. shape is not emphasized. The main interest is in system function-ality for contract design. The NEUTRABAS Planning Model that results from these considerations is shown in Fig. 50. This figure summarizes the scope of the NEUTRABAS modeling developments for ship outfitting systems.

As indicated in Section 3.1.2.4 the Ship Outfitting Systems Model from a general viewpoint consists of four main submodels, organized as independent schemas:

- Global Reference Model     - Topological and Geometrical Model
- Functional Model           - Administrative Model

These models will be discussed in more detail in the following sections. It should be noted that the Global Reference Model serves as a high level coordination framework for the ship outfitting domain to which the other three models are subordinated.

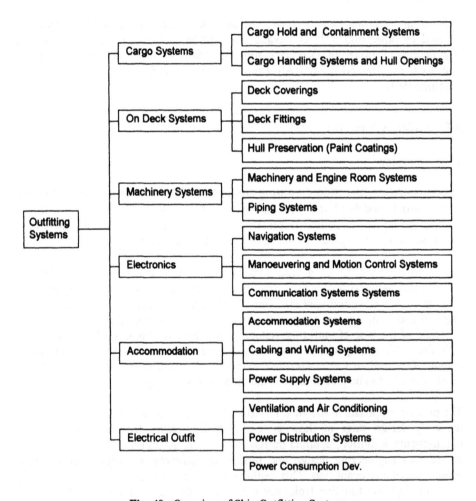

**Fig. 49** Overview of Ship Outfitting Systems

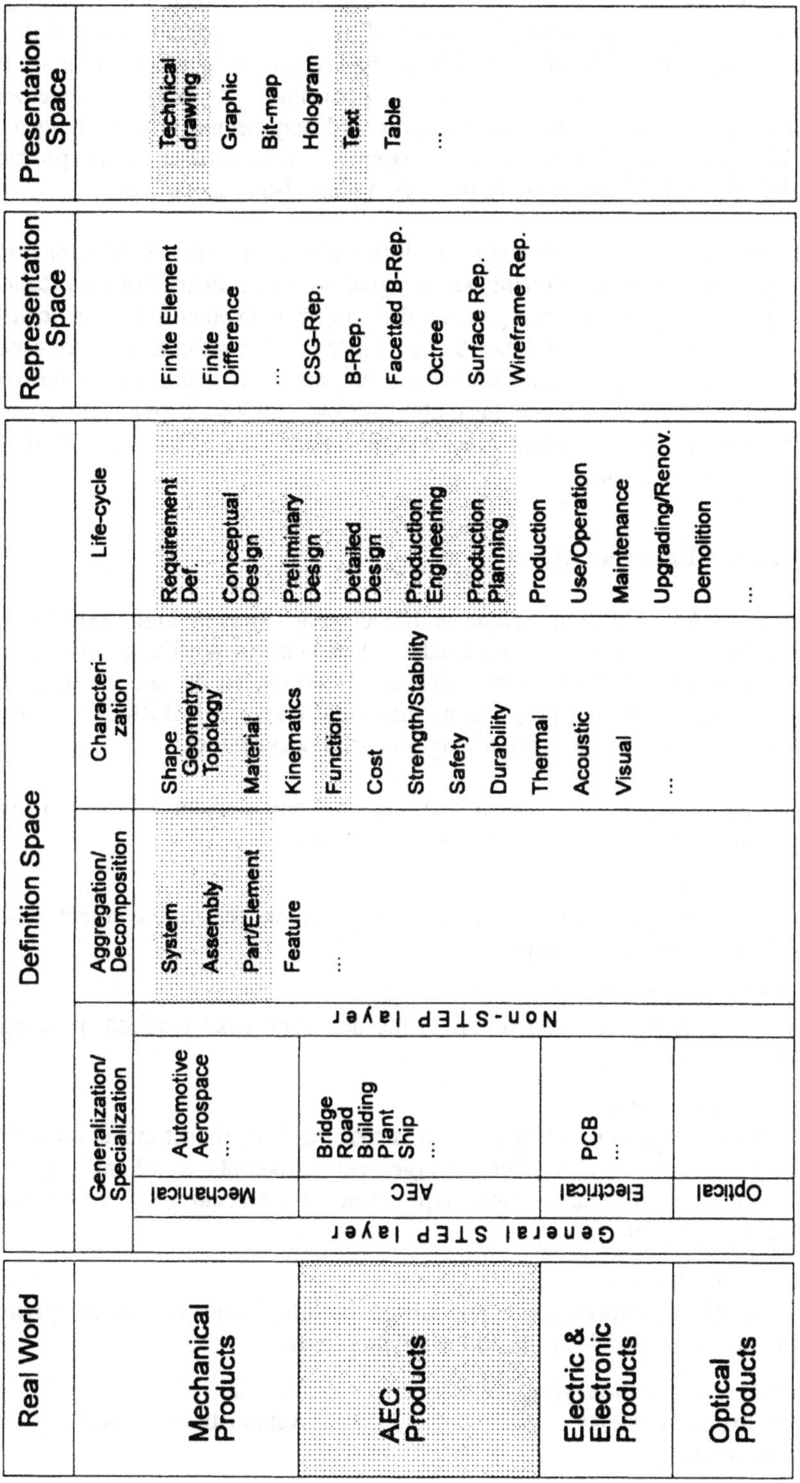

**Fig. 50** NEUTRABAS Planning Model for Ship Outfitting Systems

These models together describe both the physical properties (geometry, topology: size, location, connectivity; material etc.) of outfitting system components and their functional capabilities in the context of a system. A system is defined as a set of objects (physical or logical components) performing a function together. Components may have more than one purpose and may belong to more than one system. The scope and structure of each system must be defined in the product model. The role of each component in the system must also be identified.

A function is a role that a component or product performs in a given context or from a given viewpoint. Many components are designed for one primary function (called design purpose role). But some components also contribute to other functions (called secondary or behaviour roles) and thereby participate in other systems. E.g. hot water pipes primarily serve to distribute water, but may also play a role in the heating system. A functional product model must allow for the ability for system components to play multiple roles in different functional contexts. This is discussed in more detail in subsequent sections.

### 3.5.2 Outfitting Reference Model

The NEUTRABAS Outfitting Reference Model (Fig. 9) is a high level model describing the information structure used to document an outfitting system and serving as a reference for the product models developed for this domain. Many of the concepts used here were first proposed by the ESPRIT Project IMPPACT (ESPRIT 2165) and then adapted to the shipbuilding product model context.

In accordance with Fig. 9 a product definition for any element in the outfitting system is composed of four major sets of attribute types:

- Life-cycle stage attributes:
  Related to the life-cycle stage to which the definition pertains (contract stage, design stage, ..., production stage)

- Concretization attributes:
  Related to the degree of concreteness of the definition (like required, proposed, approved, ... , scrapped)

- Specialization attributes:
  Related to the degree to which an element is specific. The product model for a ship engine e.g. may be for a class of engines (*generic:* Two-stroke diesel engine), for a *specific* type of engine (Sulzer diesel type ...) or for the *occurrence* of an engine installed in a given ship.

- Specialization attributes:
  Related to the description of a specialized outfitting system and comprising attribute sets or submodels for the following properties:

  - Product status (feature, part, assembly, ...)    – Material
  - Geometry and topology                            – Administrative Process
  - Functionality

The refinements of this Reference Model with regard to specialization attributes are described in the following sections.

### 3.5.3 Functional Model

The Functional Model (Fig. 10) in NEUTRABAS consists of information elements in the product model that describe the functional performance of outfitting systems. The functional definition which is based on function descriptors must meet rules, laws, regulations and enterprise restrictions and be approved by various authorities.

Fig. 51 gives an overview of the system descriptor attributes. One set of these e.g. identifies the system components that belong to this system. Another set collects the function descriptors relevant to this system. NEUTRABAS has much refined the relevant attribute sets for its major outfitting systems based on the system context shown in this figure.

### 3.5.4 Topology and Geometry Model

The Topology and Geometry Model for Outfitting Systems encompasses the topological structure for these systems aboard the ship. Thus the internal connectivity of the components within each outfitting system is described, but also how these elements are embedded in the zones of the ship and are attached to the structural elements of the vessel.

Fig. 52 gives an example of how a system component is integrated into its topological context by relating it to a zone, some reference object and perhaps some assembly. The application-specific objects zone, assembly etc. are mapped onto STEP resources like region or onto concepts from other NEUTRABAS models (major surface, assembly etc.).

### 3.5.5 Administrative Model

The Administrative Model (Fig. 53) regards the outfitting systems from the viewpoint of the process in which they are involved. A process is modelled in terms of its activities which have a certain duration and are scheduled to occur at certain dates and times. The process, especially in production, consume resources, e.g., in terms of workshop space, personnel, tools etc. These attributes are planned and later executed. The activities and hence the whole process incur costs. The administrative model thus captures the relationship between elements in the product, related activities and the associated time and cost.

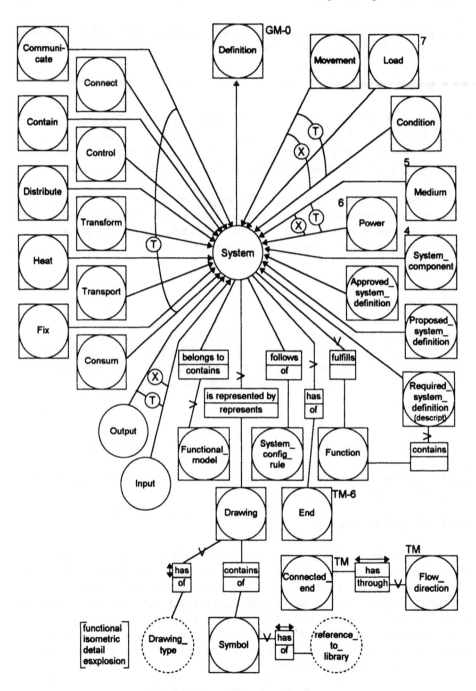

**Fig. 51** System Descriptor Attributes

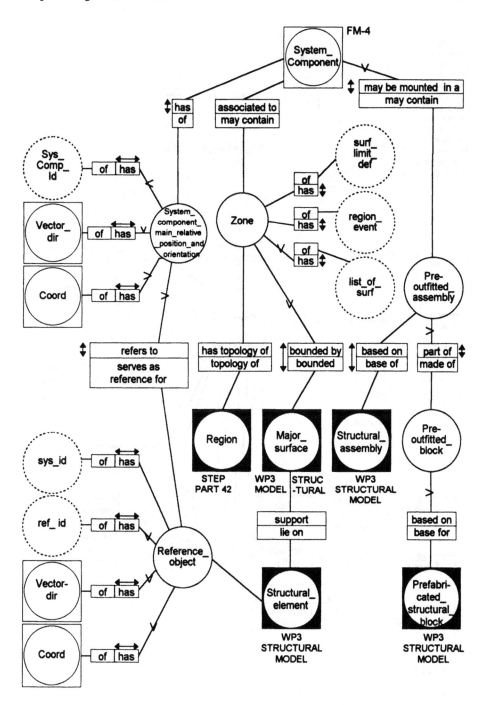

**Fig. 52**  Component General Attributes

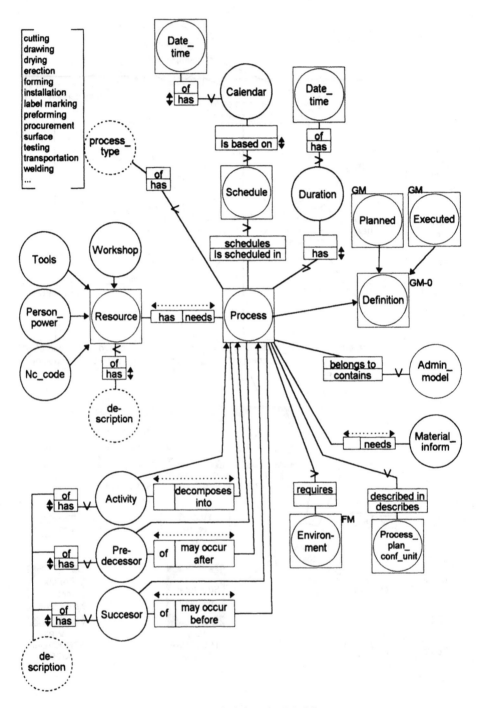

**Fig. 53** Administrative Model

### 3.5.6  Multifunctional Aspects

A product is multifunctional if more than one function is performed by the product or any of its components. A ship is a classical example of a highly multifunctional product since many of its systems serve different tasks at different times if its operation or during the life-cycle.

A multifunctional product can be evaluated from several viewpoints corresponding to its multiple roles. In NEUTRABAS each function is associated with a system and can be regarded separately. The approach is taken that distinct functionalities will be mapped onto multiple views of the database. Since components can be associated with more than one function and hence system, the database must support multiple streams of attribute inheritance, at least one for each function. This requirement certainly adds complexity to the integration of multifunctional products in a single database.

Another demanding requirement results form the fact that the functional capabilities of a product are often evaluated at several different levels of idealisation. Functional property attributes thus cannot always be assigned to the physical product model, but belong to some idealized form of the model. Pipe flow analysis, e.g., may be based on a hydraulic substitution system,  while pipe stresses are evaluated by means of some structural idealization. In consequence a multifunctional product database must administer several types of idealization of the product model. To maintain consistency simple transformation rules are desired between the physical and idealized version of the model.

In practice this complex modeling task in NEUTRABAS is approached on the basis of an object-oriented methodology. This facilitates a division of labour and a step by step procedure. The complex product is broken down into physical and functional subsets, i.e., components and systems. partial product models can be independently defined for each subset. The partial models are later integrated into a coherent product model. To facilitate the integration task, which is of course not trivial, it is a practical necessity to develop a perspective of the scope and structure of the entire product model from the beginning. This can best be documented by general application reference models.

# 4 Ship Application Reference Model

## 4.1 Approach to Integration

The approach taken in this chapter leads from the existing NEUTRABAS information models, which are separate, self-contained, topical application models, via the definition of a Shipbuilding Reference Model with a proposed schema architecture to an integrated architecture in which the role of the individual schemas is clearly identified and where the schemas are linked at specified integration points. The following steps are taken in this approach:

- The current information models are the starting point of the integration task. In NEUTRABAS, four models were developed for specific shipbuilding application areas. They are documented as follows:

  - Ship Principal Characteristics Model: Section 3.2
  - Hull Spatial Organization Model: Section 3.1.2.2
  - Ship Structural Systems Model: Section 3.4
  - Ship Outfitting Systems Model: Section 3.5

  These models, which are formally described in NIAM and EXPRESS, are self-contained for their application scope in that they contain their own application relevant schemas and references to all required STEP generic resources. In a few cases there also exist links or cross-references to other NEUTRABAS models, for example, between the Spatial Organization and Structural Systems Models. The models overlap to the extent that they share similar or identical resources, the strongest links exist in the area of ship geometry and topology which is in practice fundamental to all models.

  The form of organization of each model resembles the format of independent Topical Information Models as they were used in STEP at the time of the Tokyo Integrated Product Information Model (Tokyo IPIM, Oct. 1988). No attempt was made in these models to adopt the formal approach for the definition of Application Protocols with a clear distinction between the levels of Application Reference Models (ARM) and Application Interpreted Model (AIM) [11]. However, an organization of the NEUTRABAS models into subschemas exists so that a reorganization in an integrated schema architecture is feasible.

- The STEP approach toward integration has been much influenced by the ideas first presented by Gielingh in his General AEC Reference Model (GARM) [12]. Gielingh emphasized the need for a reference model for major application areas as a planning, coordination and integration instrument. These ideas were adopted by the STEP community and were also further developed in the ESPRIT Project IMPPACT [13, 14]. NEUTRABAS has held a close liaison with IMPPACT and has been much influenced by the reference model ideas proposed there. The shipbuilding reference model for NEUTRABAS thus aims at being compatible to the broader concepts of the IMPPACT approach.

- The STEP methodology for integration has undergone significant development and change during the NEUTRABAS project. As experiences gradually developed, the contours for a general STEP architecture and an integration approach have become clearer although there is still some flux and the corresponding documents are still draft status. These ideas gradually matured from Danner's "Proposed Framework for Product Data Modelling" [15], the "Generic Product Data Model (GPDM)" [16], the "Generic Product Data Resources for STEP" [17] to the currently valid STEP documents "STEP Development Methods" [18] and "Application Protocol Guidelines" [19]. This methodology sets a relatively well defined framework for integration. The NEUTRABAS reorganization of its schema architecture adheres to this methodology framework.

- The American NIDDESC project is developing shipbuilding product models with similar objectives, though somewhat different scope, as NEUTRABAS. A close liaison between the two projects has been maintained throughout their development. NIDDESC originally developed a Ship Structural Reference Model (Version 1.0: Dec. 87; Version 4.0, ISO Draft: June 90), which became an Application Reference Model for Ship Structural Systems submitted for STEP Part 102 [20] in June 1990. These models were similar in organizational structure to the now existing NEUTRABAS models, i. e., rather self-contained. Since January 1991 the development of a Ship Structure Application Protocol began, which has become a Working Draft, Version 0.7 as of April 1992. This new generation of the NIDDESC Ship Structural Systems Model is directed at full compliance with ISO guidelines for Application Protocols (STEP ISO Level) although the current working document is not a complete draft yet [21].

In close analogy to this history NIDDESC also developed a Reference Model for Distribution Systems (Version 1.0: August 1989) [22]. Then it proceeded to initiate developments of Application Protocols for specific shipbuilding applications of distribution systems based on this reference model. NIDDESC has submitted six proposals for Application Protocol Planning Projects which were accepted by ISO PMAG in Oslo in February 1992. They include four proposed AP projects related to ship outfitting systems:

- Ships Electrical Systems
- Ships Heating, Ventilation, Air Conditioning Design
- Ship Outfit and Furnishing
- Ship 3D Piping

These new APs will certainly also be based on STEP methodology for Application Protocol architecture. From the viewpoint of integration the NIDDESC Application Protocol set is based on two distinct reference models for ship structural systems and distribution systems. These reference models serve to coordinate two subsets of shipbuilding APs. There exists no proposal from NIDDESC for a General Ship Application Reference Model although NIDDESC is no doubt aware of the mutual interdependence between its sets of APs.

NEUTRABAS by contrast has adopted the position that the integration of all current and future shipbuilding Application Protocols requires the standardization of a single, comprehensive, high-level Ship Application Reference Model. This is a prerequisite for the interoperability of all APs and for the greatest possible extent of sharing application resource schemas among many applications.

• Against the background of these prior developments NEUTRABAS has adopted its own approach toward integration for the shipbuilding application area. NEUTRABAS defines a schema architecture which forms the framework for AP development. It consists of four levels, described in more detail in Section 2:

- – Ship Application Reference Model (with two schemas)
- – Specific Ship Application Schemas
- – Ship Integrated Application Resource Schemas
- – STEP Integrated Generic Resources

This Section defines the correspondence between the existing NEUTRABAS product models with their subschemas and the new schema architecture which is obtained by reorganization. The integration points within this new architecture will be identified. The deviations from the existing models will not necessarily be severe with respect to functionality. But the integration from one common perspective based on STEP methodology does necessitate a recasting of schemas in accordance with their role in the new architecture.

## 4.2  Reference Architecture

### 4.2.1  Integration Strategy

It is the objective of the integration process to unite the current separate NEUTRABAS information models under the roof of one single Ship Applications Reference Schema. This step will also much facilitate the integration of any future application models and the coordination at ISO level in discussions about contributions to standards from different sides.

The proposed overall schema architecture was introduced in Section 2.3 and is shown in Fig. 2. This reference architecture will also be the basis for the integration and reorganization of the NEUTRABAS schema architecture. The architecture consists of four levels as discussed earlier:

• Ship Application Reference Schema
• Specific Ship Application Schemas
• Ship Integrated Application Resource Schemas
• STEP Integrated Generic Resources

At the top level in this architecture the Ship Application Reference Schema is subdivided further, as also explained in Section 2.3 into two schemas or subschemas:

• Ship Application Reference Definition Schema,
• Ship Application Reference Representation Schema.

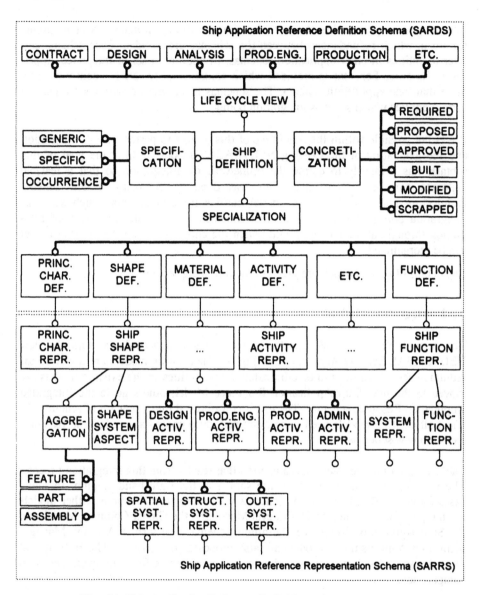

**Fig. 54** Ship Application Reference Definition Schema (SARDS)
and Ship Application Reference Representation Schema (SARRS)

Details are shown in Fig. 54. The entities at the bottom level of the SARRS in Fig.
54 are representation entities for diverse properties of the product through which the
specific application schemas at the next level inherit the product definition attributes.
Thus all applications can share these high level attribute sets of the reference model
indicated in Fig. 54.

The integration process has the aim of defining a clearly structured schema architecture in which redundancies between the schemas are avoided. Each schema can refer to entities in other application schemas. In addition a set of Integrated Application Resource Schemas is provided which contains entity sets frequently used by more than one application schema. The schemas communicate with each other via certain entities which serve as integration points.

It was not possible within the timeframe of this project to perform a full integration and reorganization of the existing NEUTRABAS models. Rather the goal of this subtask was primarily to define the framework for integration so that a feasible approach can be demonstrated. This was done by proposing a new schema architecture and identifying the crucial entities in each schema which serve as integration points with entities in other schemas. This delineates a path of integration throughout the model around which the entity sets from the existing model can be reorganized.

The details of this approach are given in the following sections.

### 4.2.2 Integration Method

The current NEUTRABAS models are reorganized in accordance with the new NEUTRABAS Schema Architecture of Fig. 2. The existing models are reviewed for redundancies which need to be eliminated. Some entities in the existing models are moved to the new Ship Application Reference Model, others go to the Integrated Application Resource schemas. The rest is consolidated into a new set of schemas. In effect all existing models need to be examined, decomposed as needed and reassembled into new schemas.

Table 1 shows the proposed reorganization that results from this integration process. The new schemas and subschemas are shown in the right-hand column. In many cases former models now become schemas with minor modifications. Other models are merged, for example, the High Level Model and the Global/Common Model into the Ship Application Reference Definition Schema. The former Activity, Management, and Administrative Models are also proposed for merging. The Hull Spatial Organization Model is decomposed into two subschemas for ship space and ship compartment.

Fig. 55 shows the new map of proposed schemas with their principal entities (integration points). The legend for the abbreviated schema names is as follows:

Reference schema:
SARRS =   Ship Application Reference Representation Schema

Specific Ship Application Schemas:
SPCS =   Ship Principal Characteristics Schema
SSAS =   Ship Spatial Arrangements Schema with subschemas for Ship Space and
         Ship Compartment

SSSS =   Ship Structural System Schema with subschemas for product, design,
         production engineering and production
SAS =    Ship Activities Schema
OSOS =   Outfitting Systems Organization Schema
SFS =    Ship Function Schema

<u>Integrated Application Resource Schemas (IARS)</u>
with possible schemas for:
– Ship product description and support
– Ship geometry and topology
– Ship product structure configuration
– Ship process description
– Ship material
– Ship function description
– Ship standard libraries


**Table 1**   Reorganization of NEUTRABAS Architecture

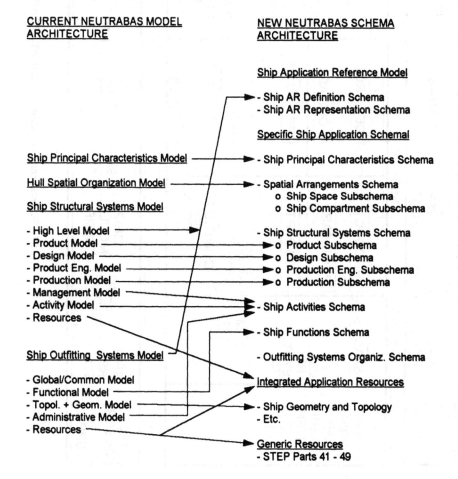

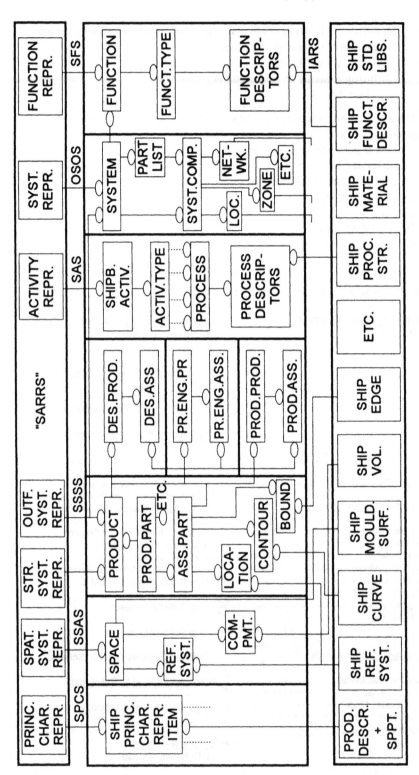

**Fig. 55**  Schemas and Integration Points in NEUTRABAS Architecture

These schemas are integrated through the following principal links:

a)   The level of the SARRS has representation entities for each of the principal specific application schemas. This representation entity is connected with a representation_item entity in its application schema. This rep_item entity is intended as a supertype or at least header entity of all other entities in the schema. This is how the application schema inherits attributes from the SARRS. The entities space, product, design_product, shipbuilding_activity, system, function etc. are regarded as representation_item entities, too, even though their name does not make this explicit.

b)   There exist several cross-connections between schemas whenever an entity uses concepts from other application schemas. For example, design_product inherits attributes from product (product subschema) and from design (SARRS). Similarly design_assembly is a special case of assembly_part as well as of design_product.

c)   The application concepts can be broken down into basic building blocks many of which can be collected as application resources. Therefore most entities at the bottom level of each application can therefore be related to resource entities in the IARS.

This integration method leads to a fully integrated schema architecture.

### 4.2.3 Adaptation of Existing Models to New Schemas

The integration process necessitates revisions of the existing NEUTRABAS models. This section describes the major steps required to reorganize the existing models and to develop the new schemas.

• **Ship Principal Characteristics Schema**
  This new schema is obtained from the existing Ship Principal Characteristics Model by renaming the top entity from ship_principal_characteristics_model into ship_principal_characteristics_representation_item.

  Further for integration and AIM development a thorough scan has to be made to determine which entities of this schema can be mapped into available STEP Generic Resources, especially from Parts 41 and 44.

• **Ship Spatial Arrangements Schema**
  The former Hull Spatial Organization Model is intended to be transformed into the new SSAS with two subschemas for Ship Space and Ship Compartmentation. In slight adaptation of Fig. 39 the space (rep_item) entity can be chosen as the header entity of this new schema. It represents the geometric and topological substance of a space and maps onto ship moulded surface. The space attribute referential_system can be related to the resource entity ship_reference_system. The compartment entity with its many attribute entities for volumes and their use can

be readily separated from the rest to form a small subschema. This entity connects with the resource entity ship_volume.

- **Ship Structural Systems Schema**
  This new schema is derived from the former SSSM. The following revisions are needed for integration:

  – The original High Level Model is replaced by the new, more general Ship Application Reference Definition Schema (Fig. 54).

  – New subschemas for Product, Design, Production Engineering, Production replace the former corresponding models. Only minor changes are required. The life cycle view attributes (design, production engineering, production) are now inherited from the SARDS.

    The different life cycle product views are integrated under assembly_part or other parts.

    The Product Subschema has the integration point assembly_part with attributes location, contour, boundary to be tied into corresponding entities at the geometric and topological resource level.

  – The former Management Model and Activity Model are proposed to be merged into the new Ship Activities Schema (see below).

- **Ship Activities Schema**
  This new schema originates by merging the former Activity, Management, and Administrative Models from Del. 3.2.2 and 4.2.2 into a single schema by generalization.

  A shipbuilding activity is interpreted in the broadest sense here. Fig. 6 from the former Activity Model gives a good variety of possible examples. The activity types in the former Management and Administrative Models can be subsumed herein or considered as special subtypes of activities.

  The entity process is used as a concept in common to all activities. A process is a set of activities. Therefore one can develop a set of process descriptors suitable for all subtypes of activities. They would for example include attributes like scheduled time, duration, cost, quality.

- **Outfitting Systems Organization Schema**
  This new schema captures only the system aspect from the predecessor model. It focusses on the system entity as top entity and describes arbitrary systems in terms of their components and part lists. These entities possess geometric and topological attributes which serve as integration points with the integrated resources.

- **Ship Function Schema**
  The SFS, which stems from the Functional Model of the SOSM, concentrates on the function description of systems. It enumerates the supported function types and provides a set of function descriptors for the whole set of types.

- **Integrated Application Resource Schemas**
At this level several new schemas can be defined for shipbuilding and shipping applications. The most important and obvious one is for ship geometry and topology. This schema will contain specific geometric and topological concepts unique to ships and more specialized and constrained than in the Generic Resources. Ship moulded geometry, ship volumes and their use are typical examples.

Other Integrated Application Resources may correspond to other areas where STEP Generic Resources exist, but where it is advisable to form a ship related, specialized subset of these. These schemas would much facilitate the definition of an AIM.

# 5 Software Developments

## 5.1 Software System Architecture

### 5.1.1 System Requirements

The representation of the ship, considered as representative of very particular complex and multifunctional products, involves the comprehensive management and exchange of sophisticated and numerous data.

In this context, the development of a database devoted to the retrieval and management of ship data, and to the adoption of efficient procedures associated with the flow of information exchange appears as a significant source of productivity improvement.

NEUTRABAS is concerned with the realization of the prototype of such a neutral database, with the definition of a standardized information model to be well adapted to the specific needs of this particular kind of a product.

The main goals for the development of a software system architecture were:

- The system should serve as a neutral format database management system for the exchange of product definition data between different CAD, KEY, CAM and CIM systems.

- It will also serve as a short and long term archiving medium for product data.

- The system will be applied in large-scale, multi-user engineering organizations.

- The system must support application processes throughout the product life cycle.

- The system must account for multifunctional products, hence must allow multiple views of the data.

- The system must be accessible by future knowledge based and expert systems.

- The database must be neutral, i.e. based on forthcoming ISO standards, open, in terms of being accessible via open networking protocols (OSI networks), and structured, that means selectively by virtue of its inherent information structure in contrast to a flat neutral file.

From the engineering viewpoint the NEUTRABAS system must meet the following major requirements :

- Support of the major activities during the CIM cycle, notably during design, production planning and production, including the relevant administrative dispositions.

- Recognition of the functional aspects of the product subsystem by accounting for attribute sets modeling the functional subsystem capabilities.

This leads to the architectural goals :

- Neutrality to the CIM-Modules
- Portability
- Dynamic Solution
- Consider Life-Cycle Aspects
- Application Modularity and Distribution Possibility
- Security

## 5.1.1.1 Neutral to the CIM-Modules

NEUTRABAS aims at supporting large multifunctional products throughout their entire life-cycle from pre-contract negotiations, through design, manufacturing and operation, up to final decommissioning.

Consequently the architecture should serve the widest possible range of CIM modules: design, production, management, and so on without bias towards any particular system or application area.

In principle, it should be feasible to develop an interface between any CIM module and NEUTRABAS.

## 5.1.1.2 Portability

The architecture must be independent of any particular hardware environment, operating system and database management system. At the same time, it employs established database management systems for data storage and manipulation.

This is achieved by a system-independent data management component, that separates applications from the specific database technology used.

Application interfaces are guaranteed independence from the underlying storage data model.

## 5.1.1.3 Dynamic Solutions

The NEUTRABAS database is used to coordinate data shared amongst a large and diverse assortment of application systems. There are potentially three ways in which applications could access data held within the neutral database.

The simplest method is off-line batch processing. Product data would be imported and exported from the database in a neutral file. This method is suitable for exchange of data between different NEUTRABAS databases, but is too inflexible for general everyday use.

A more flexible approach is to use the database as the repository of long term information throughout the product life-cycle. The database would be on-line and available for interactive access by any application at any time. To process data an application will retrieve it from the database, applying any appropriate locks. Processing of the data will take place in the applications own workspace, and finally the updated data will be returned to the database and any locks removed.

A more ambitious approach is to use the database for real time processing of product data. The database would be the source of all consistent and up-to-date product information. Applications would dynamically interrogate and update the database each time that an entity instance was required. While theoretically appealing, the real time approach would lead to many small transactions, which when coupled with the traditional access overheads of database management systems would make for unacceptably slow performance.

### 5.1.1.4 Consider Life-Cycle Aspects

The life-cycle of a ship, at twenty years, is far longer than that of most computer systems and it is essential that a NEUTRABAS model can live through several generations of database technology. The separation of the application interface and the storage database allows future extensions to the next generation of database management systems. By using EXPRESS as modeling methodology the information model becomes independent of the underlying data storage model and is portable to other database systems.

The life-cycle aspect also forces the information model to adapt to changes quickly. This leads to a late binding data manipulation language, which fetches and checks the information model when requests are processed.

### 5.1.1.5 Application Modularity and Distribution Possibility

For future extensions the NEUTRABAS system architecture must be built in a modular way, which guarantees that additional modules can be added or later existing modules can be replaced.

The modularity must also allow for separating functional blocks of the system and to use them independently. So, for example, the development and the running application can be separated on different hardware environments and each part may be run standalone.

On the other hand distribution ability is asked for. The system architecture must allow a distribution over multiple computers, connected via a network. To meet this requirement, a multi process structure must be designed.

### 5.1.1.6 Security

Regarding the situation with multiple companies using the same database, the security aspect is an absolutely necessary and important point. A security system specifying access rights for the data stored in the common data repository is needed. It must guarantee that any application may only access data it has the necessary access right for.

### 5.1.2 Functional Requirements

From the functional viewpoint the NEUTRABAS system shall support the complete cycle from the first development of an information model to the running application system.

The three main goals for the functionality of the system are :

- modeling support
- implementation support
- application systems support

### 5.1.2.1 Modeling Support

The NEUTRABAS partners have decided to use the object oriented information methodology based on the language EXPRESS.

The modeling tools of the NEUTRABAS systems shall support the development of EXPRESS schemas for NEUTRABAS as well as perform the necessary syntax checks.

An EXPRESS data dictionary is needed as a common repository for EXPRESS schemas. This allows the integration of multiple EXPRESS schemas. The EXPRESS data dictionary serves as the heart of the system.

### 5.1.2.2 Implementation Support

The information models developed inside NEUTRABAS shall be implemented on database management systems available on the market.

A data management component shall provide a database system with an independent data manipulation language, so that the applications become database independent.

A general implementation and instantiation concept for the language EXPRESS must be designed for the NEUTRABAS system. Tools for the derivation of database system dependent implementation structures as well as a database system independent EXPRESS based interface as data manipulation language are needed.

### 5.1.2.3  Application Systems Support

Applications systems shall access the data through an application program interface using the EXPRESS based interface. The application program interface must be developed once for each applications system. Having a portable system architecture and a database management system independent data access interface, the applications become portable between different installations of the NEUTRABAS system.

### 5.1.3  System Architecture

Workpackage 1 of NEUTRABAS was responsible for Information System Modeling Methodology and undertook the specification of a complete system architecture and the development of a set of basic tools to provide a test environment for the neutral data exchange between heterogeneous application systems. These tools included software for the definition and verification of EXPRESS schemas, for the generation of working space representations of EXPRESS schemas and model instances, for the administration of these working forms and for the communication with databases. The tools are EXPRESS based and application-independent. They may be interfaced with any application system.

### 5.1.3.1  Functional Architecture Blocks

The overall structure of the system architecture is given in Fig. 56. The structure reflects the main goals of the functional requirements in its functional blocks:

- modeling tools
- implementation tools
- application systems.

The modeling tools and the implementation tools share the EXPRESS data dictionary. The EXPRESS data dictionary is implemented on a database management system using the network capabilities of the DBMS and the hardware environment. This part may be installed in different environments.

The implementation tools will have a multi-process structure, which allows separating the data dictionary and the implementation storage manager.

Modeling tools and the implementation tools may also run independently when having tools to import and export EXPRESS schemas of the data dictionary. In this case, information models may be developed in one environment. Later on these models can be exported and then again be imported into the data dictionary of another implementation environment.

Application system and the implementation interface may run as different processes in different environments. This allows having multiple application-programs running at different places using one common application.

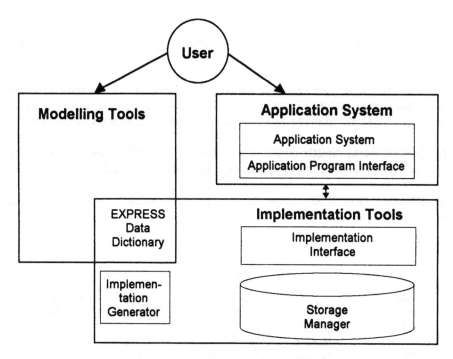

**Fig. 56**   Overall Structure of the System Architecture

## 5.1.3.2  Overview: Components of the System Architecture

Fig. 57 represents an overview of all components of the NEUTRABAS system architecture.

The modeling tools consist of :

– Operating system editor for the development of EXPRESS schemas written in an EXPRESS source file (ASCII-format)

– EXPMOD
  EXPRESS source file containing an EXPRESS schema

– EXPDD
  EXPRESS data dictionary for EXPRESS information models and the necessary administration

– EXPIMP
  Parser for EXPRESS schemas bound with the import facility to import an EXPRESS source file into the data dictionary

– EXPEXP
  Export routine extracting an EXPRESS schema from the data dictionary into an EXPRESS source file

– EXPDIC

tool for the interactive development of EXPRESS schemas on top of the data
dictionary

EXPRESS models may be developed using the EXPDIC, a tool for the interactive
development of an EXPRESS model, or by writing the EXPRESS code into a file.
The model will be stored in the data dictionary as well as the administration data.

The EXPDIC works directly on the data dictionary. EXPRESS models can also be
imported into the data dictionary by using the EXPIMP (EXPRESS import facility).

The EXPEXP (EXPRESS export facility) exports models, stored in the data
dictionary, to a file.

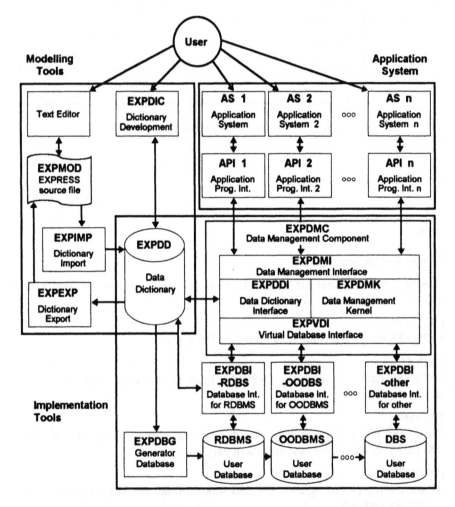

**Fig. 57**  Components of the NEUTRABAS System Architecture

The implementation tools are :

- EXPDD
  holding all information about storage managers connected, implementations and
  implementation data

- EXPDBG EXPRESS database generator deriving a database dependent implementation structure for a specified database management system

- EXPDMC database management system independent data manager representing
  an EXPRESS based data manipulation language

- EXPDBI database management system specific bridge between the data-
  management component and the storage manager

- Storage manager database management system used within NEUTRABAS for the
  storage of EXPRESS schema instances

When the development of an EXPRESS schema is finished, the database provides
the implementation structures for the chosen target storage manager. The relations
between the EXPRESS schema and the implemented structure will be stored in the
data dictionary.

There will be multiple database generators, one for each type of database management system inside the NEUTRABAS system.

The data manager (DMC) provides a schema independent data manipulation
language. It fetches all informations about an EXPRESS schema from the data
dictionary. To access the instances of the EXPRESS schema the DMC connects the
specific database interface.

The EXPRESS database interface (EXPDBI) will be implemented for each type of
database management system connected. The EXPDBI finds the information about
the mapping between the EXPRESS model and the implementation structure in the
data dictionary. It uses this information to transform the EXPRESS based calls into
database system specific requests.

The application systems consist of :

- application system

- application system interface: The application system interface connects the application system with the interface of the data management component

An application system communicates with the NEUTRABAS system through an
application interface. The application interface connects the data management
component to manipulate the EXPRESS schema instances.

## 5.2 General Concepts

The language EXPRESS is an information modeling language. There is no instantiation concept. For it's implementation NEUTRABAS has defined a general concept, how to handle instances for an EXPRESS schema. This concerns the instantiation of subtype-supertype-structures as well as security aspects like the ownership of instances and the verification of instances.

There is also a concept for the management of database systems, users, sessions and transactions.

### 5.2.1 Instances

Inside NEUTRABAS we consider schema-, subsuper- and entity-instances.

An entity-instance represents one set of data for all attributes of an entity.

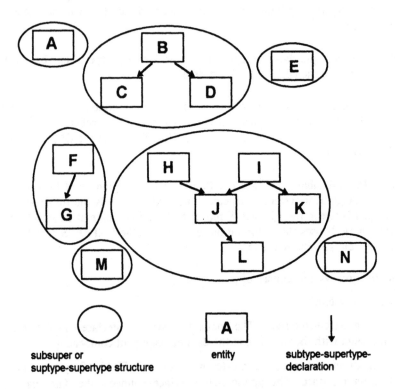

**Fig. 58**   Entities of an EXPRESS Schema with their Subtype Declarations
and the Derived Subsuper Constructs

Entities may be connected via subtype/supertype declarations (subsuper relations). The set of all entities, where any two entities have a connection via subtype/supertype declarations (using multiple entities in between) and where for each entity all supertypes and all subtypes are in this set represents a subsuper (also : subtype-supertype-structure). A subsuper instance is a set of values for the attributes of a partial set of entities of a subsuper.

Fig. 58 represents entities of an EXPRESS schema with their subtype declarations and the derived subsuper constructs.

Fig. 59 shows what entity instances and subsuper instances of this schema may look like.

A schema may have multiple instantiations. If we have for example a schema for a ship, there may be different schema instances representing the data for various ships.

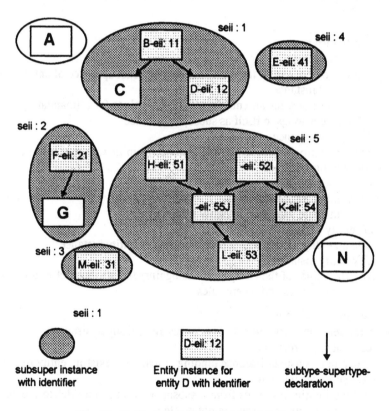

**Fig. 59**  Entity Instances and Subsuper Instances

### 5.2.2 Inheritance

For instantiation the data of all entity instances (independent or as part of a subsuper instance) are identified by a unique entity instance identification (eii). The relation between entity instances inside a subtype-supertype-structure are handled by a subsuper instance identifier (seii). For simple entities (not in a subtype-supertype-structure) the instances identified by the eii and the seii are identical.

For the internal management of entity instances storage structures will be generated with the implementation structure representing both types of instances. Each entity instance identifier belongs to exactly one subsuper instance identifier. A subsuper instance may combine a set of one to many distinct entity instances of different entities.

NEUTRABAS provides a data manipulation language (DML) for single entity-instances and subsuper-instances. Single entity-instances can be combined to sub-super-instances.

DML for entity instances :
  work on a single entity
  these functions do not care if the entity belongs to a subsuper-tree of entities
  −     create entity instance
        create an instance for an entity with at least values for allmandatory attribute
        the instance belongs to itself as subsuper instance
  −     delete entity instance
        delete one entity instanceerase this instance from its subsuper instance
  −     retrieve entity instance
        retrieve the instance of an entity
  −     modify entity instance
        modify attribute values of an entity instance
  −     copy entity instance
        create an entity instance as copy of an existing entity instance
  −     verify entity instance
        verify the values of the entity instance against the attributedefinitions, control
        uniqueness clauses and where rules

DML for subsuper relations :
  work on the relations between entity instances and subsuper instances
  −     create subsuper relation
        create a new subsuper instance binding at least two existing entity instances
  −     delete subsuper relation
        remove all entity instances from a subsuper instance and delete the subsuper
        relation the entity instances are not deleted
  −     modify subsuper relation
        add an existing entity instance to an existing subsuper instance or remove an
        entity instance from a subsuper instance the entity instance is not deleted
  −     retrieve subsuper relation

retrieve the identifier for the subsuper instance (seii) and the identifier of all entity instances (eii) of seii
- verify subsuper relation
verify the relations between the seii and all eii against the definition of the subsuper structure in the EXPRESS Information Model.

DML for subsuper instances :
work on complex subsuper instances
all parts of the subsuper instance are identified by one subsuper instance identifier
- create subsuper instance
create all entity instances for a subsuper instance or create a subtype and all its supertype instances for a subsuper instance (special case : create the entity instance for the root of the subsuper tree)
- delete subsuper instance
delete all entity instances for a subsuper instance or delete a supertype and all its subtype instances
- retrieve subsuper instance
retrieve all entity instances for a subsuper instance or retrieve a subtype instance and all its supertype instances
- modify subsuper instance
add an entity instance to an existing subsuper instance or remove an entity instance from a subsuper instance
- copy subsuper instance
copy all entity instances of the subsuper instance or copy a subtype instance and all its supertype instances
- verify subsuper instance
verification of all entity instances in the subsuper-tree verification of the sub-super-structure (existence of entity instances, abstract supertype,supertype expression).

The following example will illustrate the functionality of the data manipulation language (DML):

```
...
type
   persno        = member;
end_type;

entity person
   supertype of (employee andor participant andor
trainer);
     first_name   : name;
     last_name    : name;
     address      : text;
unique
     person_id    : first_name, last_name;
end_entity;
```

```
entity employee
     subtype of (person);
     persno         : persno;
unique
     employee_id   : persno;
end_entity;

entity participant
   subtype of (person);
   of_course     : set [1 : ?] of course;
end_entity;

entity trainer
   subtype of (person);
   pref_town     : optional set [1 : ?] of name;
end_entity;
```

. . .

The example is represented in Fig. 60.

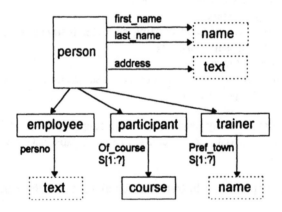

**Fig. 60**  Illustration of the Functionality of the Data Manipulation Language (DML)

A person who is an employee has an entity instance in person and one in employee. Both are identified by the same subsuper identification (seii). If this person becomes a participant an entity instance for participant has to be generated with the same eii :

create     subsuper instance with entity instances for person and employee
create     entity instance for participant
modify     subsuper relation add entity instance for participant.

If this person is no longer an employee :

modify     subsuper instance remove entity instance for employee
delete     entity instance for employee.

When we are no longer interested in the person :

delete      subsuper instance for person.

This includes the deletion of all subtype entity instances of person.

Another problem we have to solve is represented in Fig. 61

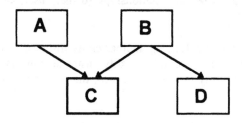

**Fig. 61**  Subsuper Entity Dependence

```
entity A;
  col_a_1 : ...
  ...
end_entity;
entity B;
  col_b_1 : ...
  ...
end_entity;
entity C
  subtype of (A,B);
  col_c_1 : ...
  ...
end_entity;
entity D
  subtype of (B);
  col_d_1 : ...
  ...
end_entity;
```

Let us imagine we have already created an instance for A (eii_A) and an instance for B (eii_B) and now we want to create an instance for C, which belongs to eii_A and eii_B. At this moment eii_A, eii_B and the instance for C should be treated as one instance.

The different steps of this example are taken by the following commands :

create      entity instance for A -> eii_A create
            entity instance for B -> eii_B

create        subsuper relation for eii_A and eii_B -> seii_AB

create        entity instance for C -> eii_C

modify        subsuper relation for seii_AB add eii_C.

### 5.2.3  Ownership / Usage of Instances

In this chapter we describe the relationships (owner, user, null) between entity instances.

User is an entity instance referencing another entity instance. Owner is a user, who owns the referenced entity. No other entity instance may use an owned entity instance. If the owner entity instance is deleted, the owned entity instance will also be deleted.

To control the ownership in the relational database we use an entity ENTITY_OWNER, where we may store in which schema instance which entity instance is the owner of another entity instance. If the owner is a group, we find all user instances of the entity instance in the entity ENTITY_USER.

Let us explain the meaning of the ownership with an example :

```
...
entity line;
   start    : point;
   end      : point;
end_entity;
entity point;
   x        : real;
   y        : real;
   z        : real;
end_entity;
...
```

In the implementation we find in the attributes START and END of entity LINE referring (Entity-Reference) to entity POINT.

Points can appear here in three forms : independently, used or owned. That means :

- independently : a point may exist standalone, meaning that there exist no references to that instance of point
- owner :         a point may be owned by one line, no other line use or own this point.
- used :          a point may be used by multiple lines, no line can be owner of this point, owner of this point is a group of lines using this point

In general :

- an entity instance may be owned by 0:1 entity instances.
  an entity instance may own 0:* entity instances.

- an entity instance may be used by 0:* entity instances.
  an entity instance may use 0:* entity instances.

  If an entity instance is owned by another entity instance no other entity instance may use it. If an entity instance is used by more than one other entity instance none of them may own it.

To control these rules we introduce the following definition of an owner :

Owner is one entity instance or a group of entity instances using this point.

   A dependent entity instance is dependent of 1:1 owner.

   An owner consists of 1:* entity instances ownership:
   the owner consists of 1:1 entity instances usage :
   the owner is a group consisting of 1:* entity instances

   An owner may own 1:* entity instances.

If an entity instance with references to other entity instances is created, deleted or modified, ENTITY_OWNER and/or ENTITY_USER need to be modified.

In all functions when working on an attribute with an entity reference we may add the reference mode : usage or ownership. The default is usage. In addition there is a set of functions to work on owner-instances.

We work on the data with the following rules :

- No entity instance can be created until all (not optional) referenced entity instances have been created.

- An entity instance may be used by multiple other entity instances (owner = user group) or owned by only one instance (excluding other usage).

- No entity instance can be deleted as long as an owner or user of this entity exists.

- Deletion of an entity instance includes automatic deletion of all owned entity instances (not used entity instances).

## 5.2.4 DBS-Sessions, Transactions and Verifications

### 5.2.4.1 DBS-Sessions

DBS-Sessions are related to one storage manager and to one schema instance. After connecting to the database system, the application must open a schema instance to manipulate its data. Before leaving the system, the application has to close the schema instance.

### 5.2.4.2 Transactions

We would need a general transaction-concept at the level of the data management component to allow consistent work with multiple connected storage managers.

For NEUTRABAS we had to simplify the concept. Multiple users may work at the same time on different schema instances. At one time only one user may change data in the schema instance. Other users may access this specific schema instance only to read data. To manipulate instances (create, delete, modify, copy) a transaction must be opened.

The actual state of a schema instances is stored in the data dictionary.

A rollback mechanism resets all changes done since the start of the transaction (undo) and sets the access to read.

The commit mechanism verifies all changes made since the start of the transaction. If there are no contradictions with the EXPRESS schema the changes are fixed in the database system and the data dictionary.

### 5.2.4.3 Verification

Verifications may be done for :

* schema instance
  verification of
         entity instances
         subsuper instances
         subsuper relations
         rules
  results in SCHEMA_VERIFY

* subsuper instance
  verification of
         entity instances
         unique-clauses for multiple entities of the subsuper tree
         where-clauses for multiple entities of the subsuper tree
         subsuper relations
  results in SUBSUPER_VERIFY

* entity instance
  verification of
         attributes
         where-rules
         unique-clauses
  results in ENTITY_VERIFY

- subsuper relation
  verification of
          existence of entity instances
          supertype-expressions

- transaction
  verification of all changes done since start_transaction

  storage depends on changes

- rule
  verification of rule definition for existing entity instances.

Every modification of a subsuper or an entity instance sets the verification status of this instance and of the schema instance to open. A modification of a subsuper relation will also be treated as a modification of the subsuper instance.

The verification of instances can be invoked by the application for each type of instance. When committing a transaction the verification state of all instances will be checked and if necessary the verification will be started.

### 5.2.4.4 Administration of User and Database System (DBS)

To administer the access for users to schema-instantiations a general user-concept on the level of the data dictionary is needed.

Users are identified by their name and password. They may have different access-rights to the data dictionary. The user may be generator of the DBS, creator of schema instances or he may have different access-rights to schema instances. Only a user with access rights to the data dictionary may generate schemas, create schema instances or work on schema instances.

The entity DBS_USER in the data dictionary stores all information about users. The ownership of generated schemas may be found in the entity DBS. The ownership of schema instances is stored in the entity SCHEMA_INST.

The access-rights for DD-user are :

s   system
    may create new user
    includes access g

g   generator
    may generate schemas
    when generating a schema the information will be added to the table DBS with
    the reference to the user in DBS_UID
    includes i

i   instantiator
    may create schema instances

when creating (create_schema) a schema the information will be added to the
table SCHEMA_INST with the reference to the user in dbs_uid
includes a

a   access
access to a schema instance
working on a schema instance depends on the given access-rights stored in the
table in DBS_ACC
access-rights may be given to a user from an other user with grant_dbs_acc.

To handle users and their access rigths we need the functionality :

- connect data dictionary
- disconnect data dictionary..
- grant user
- revoke user
  including the possibility to alter the ownership of the DBS and/or schema
  instances if exist
- alter user, they alter password or access-right including the possibility to alter
  the ownership of DBS and/or schema instances if they exist
- describe user including the ownership of DBS and schema instances.

Two users will be created automatically :

system/<pwd> acc=s                      public/public acc=a.


## 5.2.4.5  Access-Rights to Database Systems

Only a data dictionary user may have access-rights to implementations. The access-
rights to database systems are :

r - read                              w - write (includes read).

The level for the access rights are :

schema        spi        schema instance    sii      entity-instance    eii.

The owner of the schema generation has the write access to all schema instances.

The owner of the schema instance has the write access to all entity instances created
in this schema instance.

There is a hierarchy for the different access-rights

| Hierarchy | spi | sii | eii |
|-----------|-----|-----|-----|
| 3         | x   | -   | -   |
| 2         | x   | x   | -   |
| 1         | x   | x   | x   |

which means that the access-rights include those of the next level below.

## 5.3 Software Components

The scope of the EXPRESS based tools implemented by NEUTRABAS corresponds to the specification given in the preceding section except for minor omissions in details of functionality, which were not required for the feasibility tests and demonstrations performed by this project. The structure of the implemented system is in full agreement with the architecture described above.

The implementation is based on the following system environment:

| | | |
|---|---|---|
| Operating system | : | UNIX |
| Programming language | : | C (ANSI), GNU-C-Compiler recommended (GCC) |
| Database system | : | ORACLE Version 5 with SQL*Forms Version 2.3, RPT and PRO*C precompiler |

All EXPRESS based tools are based on the EXPRESS language definition of July 1990 (N 64) document. It was necessary for the project to adopt a fixed version of EXPRESS for its pilot developments because these had to be begun and performed long before the final release of EXPRESS for DIS balloting (May 1991).

Except for the EXPIMP, based on the NIST tools, the tools are updated to the EXPRESS version (N14) for DIS balloting.

### 5.3.1 Modeling Tools

The main goal of the modelling tools is the support of the development of the EXPRESS schemas. They will also do the necessary syntax check. EXPRESS schemas can be developed and stored as ASCII file or inside a common EXPRESS data dictionary. Import and export tools provide the transformation between written EXPRESS schemas in a file and schemas stored in the data dictionary.

Fig. 62 represents the modeling tools of NEUTRABAS.

### 5.3.1.1 EXPMOD : EXPRESS Source File

An EXPRESS source file is an ASCII file containing the EXPRESS code for an EXPRESS schema. The file can be developed using any editor available on the operating system.

Using the EXPRESS import an EXPRESS schema can be loaded into the EXPRESS data dictionary. The EXPRESS export allows to extract an EXPRESS schema from the data dictionary into an EXPRESS source file.

EXPRESS schemas can be exchanged between multiple NEUTRABAS systems via the EXPRESS source file.

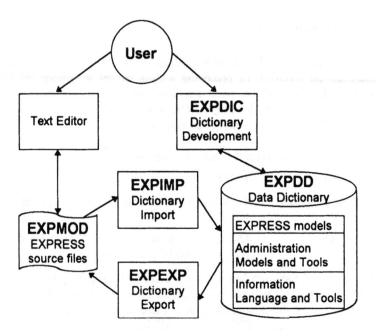

**Fig. 62**    The Modeling Tools of NEUTRABAS

### 5.3.1.2 EXPDD : EXPRESS Data Dictionary

The data dictionary is the kernel of the NEUTRABAS system. It stores all EXPRESS schemas developed inside the system and provides this information to the implementation tools (database generator and data management component). Additionaly,the data dictionary contains the administrative data for the implementation schema.

The NEUTRABAS data dictionary is divided into the following parts :

• EXPMODEL
All aspects of the language EXPRESS (Status : ISO N14) can be stored inside EXPMODEL.

An EXPRESS schema can be imported into EXPMODEL or it can be developed interactively using the data dictionary tool. The export provides the extraction of an EXPRESS schema into an EXPRESS sourcefile.

The data management component uses the EXPMODEL information for the validation of instances for an EXPRESS schema.

The constructs of the language EXPRESS are represented in storage structures for :

o   schema
o   parts
        . interface
        . constant

. type
. entity
. (incl. supertype expression and subtype clause)
. function
. procedure
. rule
o components
   . attribute
   . where rule
   . unique clause
   . parameter
   . function result
   . local variables
o expression
o procedural code
o remarks and documentation

- EXPADMIN
EXPADMIN provides the necessary administration for data dictionary user and EXPRESS schemas.

All tools accessing EXPRESS schemas inside the data dictionary are using the EXPADMIN.

EXPADMIN allows composing projects of multiple EXPRESS schemas. Regarding the life cycle aspect projects are versioned. Projects have a data dictionary user as owner who may grant access to other data dictionary user.

- EXPLANG
EXPLANG provides the base topics of the language EXPRESS, as there are:

o reserved words
o builtin functions
o builtin procedures
o builtin constants
o ...

EXPLANG may be used by all tools providing help for the development of EXPRESS schemas.

- EXPTOOL
EXPTOOL provides the information needed for the EXPRESS dictionary development tool. These are tool menues, schema views etc..

- EXPDBS
EXPDBS is the administration part needed for implementations of EXPRESS schemas inside a database management system and their instances.

The data management component uses this part of the data dictionary to access EXPRESS schema instances inside a database system.

It provides the information :

- o how to access a specific database management system inside the NEUTRABAS system
- o on which database system an EXPRESS schema is implemented
- o users and their access rights for schema instances
- o status of instances and their verification

- EXPRDBS
EXPRDBS represents the relations between an EXPRESS schema and an implementation structure derived for a relational database management system.

After a successful completion of a database generation for an EXPRESS schema the database generator stores the results of the generation process into the EXPRDBS part of the data dictionary.

The data management component uses this information for the access of EXPRESS schema instances inside the database system specific implementation structures.

- EXPOODBS
EXPOODBS represents the relations between an EXPRESS schema and an implementation structure derived for an object oriented database system.

The database is realized based of ORACLE using SQL command procedures to administer the EXPDD. Based on ORACLE, the EXPDD can be installed on any operating system where ORACLE is available.

### 5.3.1.3  EXPDIC : EXPRESS Dictionary Tool

The EXPRESS Dictionary Tool EXPDIC serves for the development and maintenance of EXPRESS information models stored in the EXPDD as well as common user interface to all EXPRESS based tools.

As common user interface the dictionary tool provides access to :

- EXPRESS schema browser and development
The user may choose an EXPRESS schema and invoke the browser and development tool for it.

- EXPRESS administration
A user can maintain the access rights for his EXPRESS schemas stored inside the data dictionary.

- Data dictionary administration
The database administrator for the NEUTRABAS EXPRESS data dictionary maintains the data dictionary user and the access rigths to the data dictionary.

- EXPRESS export
An EXPRESS schema can be chosen from the data dictionary and the export into an EXPRESS source file can be started.

- EXPRESS import
  The user may invoke the EXPRESS import facility for a given EXPRESS source file.

- EXPRESS database generator
  The user may choose an EXPRESS schema stored inside the data dictionary and start the generation process for the implementation structures for a given target database management system.

The browser and development tool provides :

- information on EXPRESS schemas stored inside the data dictionary.

  The user can access schemas, parts and components directly, he may scroll through lists of EXPRESS constructs or he may follow the EXPRESS structure. Starting with an overview of all constructs of an EXPRESS schema the user may choose and directly access the desired object. So, he may search through an EXPRESS schema to find the object he wants to get information about.

- modification of EXPRESS schemas.

  All information found and displayed can be changed interactively. The user may add new EXPRESS constructs or delete existing constructs. All modifications can be committed or rolled back by the user. For all modifications the main rules and constraints of the language EXPRESS are checked by the development tool.

The EXPRESS dictionary is implemented using the ORACLE tool SQL*Forms 2.3.. So it is bound to the ORACLE database management system. It is portable to all hardware environments and operating systems where the ORACLE DBMS is available.

### 5.3.1.4 EXPIMP : EXPRESS Import

The EXPRESS import tool imports an EXPRESS schema stored in an EXPRESS source file into the data dictionary. The access rights will be set for the user who starts the import and can be maintained using the EXPDIC tool.

The implementation is based on the NIST EXPRESS Working Form with its associated EXPRESS parser, Fed-X. This software is a set of public domain software tools for manipulating information models written in EXPRESS. This toolkit consists of a set of C language procedures and data structures. A parser reads EXPRESS source files and creates the Working Form representation of the model. This toolkit is integrated into the NEUTRABAS tool EXPIMP.

The NEUTRABAS import is a C program reading the EXPRESS schema definitions from the NIST toolkit and writing these informations by using embedded SQL to the data dictionary. The tool is restricted to the hardware environments and operating systems supported by the NIST parser.

The NIST toolkit is written in ANSI standard C. It is based on the July 1990 version of the EXPRESS language (Doc. N 64). Because of the restriction to UNIX like operating system, the EXPIMP is restricted to these systems, too.

### 5.3.1.5 EXPEXP : EXPRESS Export

The EXPRESS export extracts information for an EXPRESS schema from the data dictionary and writes the derived EXPRESS code into an EXPRESS source file (ASCII format). The export tool checks if the user, who start the export, has the necessary access-rights to the schema to be exported.

When exporting, the export tool orders the components of a schema by their type (interface, constant, type, entity, function, procedure, rule) and inside one component type by their name. This provides of long schemas a good readability for the reviewer.

The export tool is implemented using the ORACLE tool RPT, which is an SQL based report writer tool. It is portable to any system where the ORACLE DBMS may run.

### 5.3.2 Implementation Tools

### 5.3.2.1 EXPDBG : EXPRESS Database Generator

The EXPRESS Database Generator produces a database system specific implementation structure for a given EXPRESS schema. The generator takes the input schema from the data dictionary.

During the specification process, two approaches were examined :

• mapping of an EXPRESS schema onto a semantically irreducible information model and start the derivation of the implementation structure with the semantically irreducible information model

• derivation of the implementation structure directly from the EXPRESS schema.

The first approach contains a validation of the EXPRESS schema. Following this approach, the semantically irreducible information model would be the same for all target implementation databases. Only the second step would be database system dependent. This approach contains a validation of the EXPRESS schema.

The second approach takes less effort but it also contains less validation. The EXPRESS schema is taken as it is and directly mapped to an implementation structure.

For the test implementation NEUTRABAS decided to implement the second approach, called the "simple version" of the database generator.

The EXPRESS Database Generator has to be specified and implemented for each type of database management system used as storage manager inside the NEUTRABAS system.

For the target test purposes of NEUTRABAS, only the relational database generator (EXPRDBG) is specified and implemented. Others can be added in a modular way.

The information flow for the database generator is represented in Fig. 63.

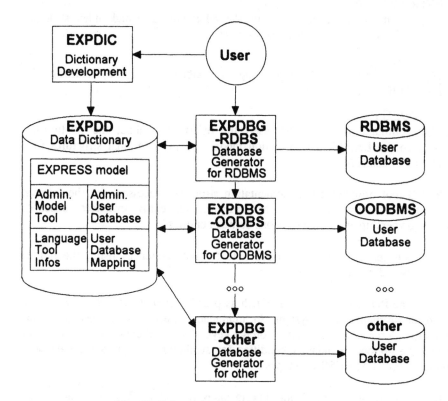

**Fig. 63**  Information Flow for the Database Generator

### 5.3.2.2  EXPRDBG : EXPRESS Relational Database Generator

The EXPRESS Relational Database Generator Tool EXPRDBG was implemented as a simple version. For an EXPRESS schema stored in the EXPDD the relational structure (implementation schema) is derived. An SQL command procedure is provided to build the implementation in a user database.

The mapping relations between the EXPRESS schema and the derived implementation schema are also provided in a SQL command procedure, which should be started in the EXPDD to fill the EXPDBS/EXPRDBS part.

The mapping information provides the necessary information where to find the data of the EXPRESS schema for the schema independent interface .

The created output files are:

–  EXPRDBG.CRE
   SQL command procedure to create the relational table and index structure derived for an EXPRESS information model.

–  EXPRDBG.DRO
   SQL command procedure to drop the tables created by starting the EXPRDBG.CRE file.

–  EXPRDBG.INS
   SQL command procedure to insert the information about the relationship between the EXPRESS information model and the relational structures in a relational database.

The RDBG tool provides all implementation structures necessary for the internal schema administration. They are the structures for entity instances, subsuper instances, ownership and usage relations between entities.

Relations between entities are mapped onto foreign references using the unique entity instance identifier (eii).

In general the RDBG tries to use as much as possible available functionality of the target database management system. So, uniqueness clauses are mapped to unique keys, entity references are mapped to foreign keys. Relational database systems allow mapping the optional/mandatory condition for attributes as well as the data definition for simple EXPRESS data types.

EXPRESS entities are mapped onto one to many relational tables. For each entity one main table will be generated. Depending onto the definition attributes are mapped to columns or additional tables.

Especially attributes defined as aggregations require additional tables, which have a foreign reference to the main entity table and an index for the aggregation values. Aggregation chains with uniqueness conditions inside may require generating multiple tables. Enumerations are mapped onto one table, which may be referenced from multiple tables.

The relational database generator tool is implemented using the ORACLE tool RPT. It is portable everywhere the ORACLE DBMS may run.

### 5.3.2.3 EXPDMC : EXPRESS Data Management Component

The data management component bridges the gap between the application interface (API), the EXPRESS data dictionary (EXPDD) and the database system interface (DBI).

Fig. 64 represents the data management component and its relations with the data dictionary, the database interface and the storage manager.

It manages the whole administration of database systems, works on the consistence between real data and the EXPRESS information model and operates on the stored data.

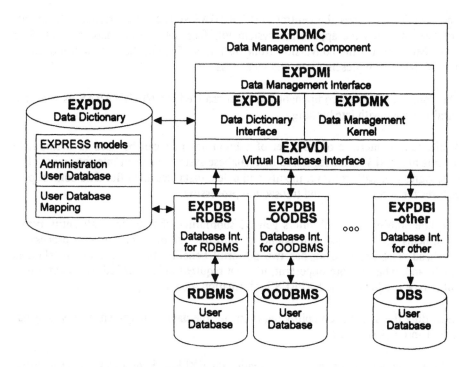

**Fig. 64** The Data Management Component

The following validations have to be performed at the EXPDMC level :

– supertype-expressions for entities

– existence of entity attributes, if they are not optional

– definition of aggregation-chains

– unique clauses of entities
(this is needed if later on the DMC shall work in distributed database systems -
for the purpose of NEUTRABAS working on a relational database system, this
control will be done automatically by the database system)

– where-rules of entities
(the EXPRESS data dictionary for NEUTRABAS does not decompose
expressions - so this part has not been realized for NEUTRABAS)

– rules defined in the schema
(the EXPRESS data dictionary for NEUTRABAS does not decompose procedural
code - so this part has not been realized for NEUTRABAS)

– dependencies between entities.

The EXPDMC is thus the central NEUTRABAS software tool responsible for all
processes between the application system interface and the database systems. The
EXPDMC controls the conversion from instances of application product models to
neutral database representations and vice versa.

The data management component has been realized for NEUTRABAS as a procedural interface with late binding.

A procedural interface offers a set of subroutines that can be used to manipulate
schema and entity instances stored in a database system. Such an interface must offer
procedures to create, retrieve, modify and delete instances and offer data control procedures for access control and transaction processing.

The late binding approach offers generic subroutines that apply to all entity types.
Arguments are validated and bound at run-time, adding considerable processing
overhead. The late binding has two powerful benefits: the procedural interface is
small, and, what is more important, it is not required to be changed whenever a new
entity type is added to the EXPRESS information model.

Regarding the distribution ability requirement, the data management component has
a multi-process structure.

The data management component provides for EXPRESS instances a data manipulation language for simple entity instances, subsuper instances and subsuper relations
between instances. A set of functions working on the ownership and usage between
entity instances is also available.

The following list presents the functionality of the DMC :

Meta-DCL for DD-Sessions
Connect Data Dictionary
Disconnect Data Dictionary

Meta-DCL for DD-user
Grant User
Revoke User
Alter User
Describe User

DCL for DBS Access-Rights
Grant Access-Rights to
Instanciation
Revoke Access-Right to
Instanciation
Alter Access-Right to Instanciation
Verify Access-Right to Instanciation
Describe Access-Right to
Instanciation

DDL for EXPRESS-Model
Describe Schema
Describe Entity
Describe Subsuper
Describe Attribute
Describe Where-Rule
Describe Unique-Clause
Describe Rule

DML for Schema Instances
Create Schema Instance
Drop Schema Instance
Verify Schema Instance

DML for DBS-Sessions
Open Schema Instance
Close Schema Instance

DML for Transactions
Start Transaction
Commit Transaction
Rollback Transaction
Verify Transaction

DML for Entity Instances
Create Entity Instance
Delete Entity Instance
Retrieve Entity Instance
Modify Entity Instance
Copy Entity Instance
Verify Entity Instance

DML for Relations between Entity
Instances
Create Owner
Delete Owner
Modify Owner
Retrieve Owner

DML for Subsuper Instances
Create Subsuper Instance
Delete Subsuper Instance
Retrieve Subsuper Instance
Modify Subsuper Instance
Copy Subsuper Instance
Verify Subsuper Instance

DML for Subsuper Relations
Create Subsuper Relation
Delete Subsuper Relation
Retrieve Subsuper Relation
Modify Subsuper Relation
Verify Subsuper Relation

The EXPDMC consists of four main modules:

- EXPDMI

  The Data Management Interface (DMI) examines the conditions for performing a function call on the DMC. A valid call is passed on the DMK.

- EXPDMK

  The Data Management Kernel (DMK) provides all administrative and technical functions to manage and instantiate the database. It makes use of the EXPDDI for interrogation of the Data Dictionary. Valid function calls from EXPRESS models are decomposed into primitive function calls for execution by the EXPVDI/EXPDBI and EXPDDI.

- EXPDDI

  The Data Dictionary Interface (DDI) handles all queries from the DMK to the EXPDD concerning the underlying EXPRESS schema and the administrative data.

- EXPVDI

  The Virtual Database Interface (VDI) bridges the gap between the EXPDMC and the EXPDBI, which depends on the database type of the user database for the implementation.

The Data Management Component is based on the Pro*C interface to ORACLE and implemented in C.


### 5.3.2.4  EXPDBI : EXPRESS Database Interface

The EXPRESS Database Interface (DBI) organizes the step from a database type independent EXPRESS model function call to the function calls based on the implementation schema.

The EXPDBI is a database system dependent set of procedures called by the DMC to manipulate data in the database (create, delete, retrieve, modify and verify). The DBI also provides data control functions to connect and disconnect the database and to manage transactions.

The DBI has two principal parts:

- The Data Dictionary part which fetches the mapping from the EXPRESS model to the relational structure.

- The Database System part which processes DMC requests and translates them into operations on the specific database.

In NEUTRABAS the DBI was implemented only once, namely for the relational database system ORACLE. It is implemented in ANSI C with SQL using the ORACLE Pro*C precompiler version 1.3. The EXPRESS based DML is translated into dynamic SQL statements.

The implemented form of the DBI for ORACLE is called EXPDBI-RDBS. It translates DMC queries into dynamic SQL statements.

## 5.4 Lessons Learned

The specification of the DMC showed that for the administration of DBS we need the information on the existing entity instances and the relationships between them. We also need this information at the DBS. The solution of this problem could be a third level database for the administration of DBS including the administration of instances. Then we would need to have unique identifiers for instances in this database.

In effect we would need three levels of databases :

EXPDD    – data dictionary for the administration and storage of EXPRESS Information Models
The new EXPDD would consist of the parts EXPADMIN, EXPMODEL and EXPTOOL of the od EXPDD

EXPINS    – administration of database systems and instances
The EXPINS would consist of the parts EXPDBS of the old EXPDD and the tables ENTITY_OWNER, ENTITY_USER and SUBSUPER of the old RDBS

DBS    – multiple database systems for the storage of instances
The tables ENTITY_OWNER, ENTITY_USER and SUBSUPER would be dropped from the DBS.

For NEUTRABAS most of the specification work had already been done and it would have taken too much effort to change it. In the two level approach (EXPDD and DBS) for NEUTRABAS we consequently have to store this information at both levels, which is not a perfect but a feasible way.

## 5.5 Demonstration and Test Cases

Early in the NEUTRABAS project a structural MOCK-UP was identified as the prototype on which to base subsequent analyses, implementations, testing and demonstrations.

In choosing the MOCK-UP care was taken to ensure that the type of ship selected and the portion of the ship to be considered should be a good representation of a complete ship in terms of the number of entities envisaged at that time.

The structure situated at the fore end of a container ship was chosen because its arrangement covered most of what the experts on the project could foresee in terms of complexity.

Bounded by the curved hull of the ship, two decks and encompassing longitudinal and transverse bulkheads, these major surfaces or "primary structure elements" formed a number of compartments. The primary structure elements themselves were composed of plates, plate sheets and assemblies reinforced with welded stiffener elements. The structure also contained a variety of apertures and cutouts for stiffener penetrations and access for people. The whole mock-up was also capable of being broken down into more than one "prefabricated block" for construction purposes.

In addition to the structural steelwork contained within the mock-up, and in order to cover a wider range of applications, the complete hull form of the container ship together with a number of internal surfaces outside the range of the mock-up were modelled.

In establishing which applications to use in the tests the following considerations were taken into account:

• The need to simulate as many phases in the life cycle as possible in the time scale available.

• The applications, whilst deliberately chosen to address different problem areas, should address and access a number of common entities.

• As many as possible of the entities and attributes defined in the MOCK-UP schema should be utilised.

• The tests should demonstrate NEUTRABAS' capabilities both as an on-line dynamic data repository and as a medium for product data exchange.

• The test should fully validate the architecture of the NEUTRABAS system software.

With these considerations in mind 4 applications were chosen covering 4 life cycle phases as follows:

    PAN - Conceptual/Preliminary Design (Spain)
    SLAS - Preliminary Design and Contract Design (Spain)
    CADIS - Detail Design (UK)
    CRESTA - Production Planning (France)

The choice of these systems had the added advantage of demonstrating the "openness" of the NEUTRABAS system. PAN, SLAS and CRESTA all operate on DOS based PCs whilst CADIS is a mainframe based CAD/CAM system.

Ideally all four applications should have linked to a single UNIX based NEUTRABAS database. In practice however this proved to be too difficult to organise simply because the interface development work was carried out in Spain, France and the UK, with little time left to integrate. Nevertheless it was possible to link the CADIS and CRESTA applications to the same UNIX NEUTRABAS database. PAN and SLAS communicated with the same PC version of NEUTRABAS.

Care was taken to ensure that the Data Dictionary and Schemas were identical on both versions.

Each module of the NEUTRABAS system, ie. the Data Dictionary, the EXPIMP and EXPEXP modules, the Data Management Component, the Relational Database Interface etc. were to be involved in the tests. In particular as many of the functions of the Data Management Component were to be tested as possible.

### 5.5.1 The PAN and SLAS Interfaces

The PAN system is a set of modules used to calculate important characteristics of a ship's hull form which are crucial to the naval architect, such as hydrostatics, stability, powering, etc. These calculations are carried out at an early stage in the design process to establish that the hull form shape possesses the desired characteristics which make it fit for purpose.

More often than not these calculations are carried out long before the final "faired" hull form has been determined and as such operate on "preliminary" hull form geometry defined by a set of points in 3 dimensional space. Only when the naval architect is satisfied with this preliminary form would the final fairing process begin.

The SLAS system takes the preliminary form of PAN and carries out the fairing process to produce the final hull form to which the ship will be built. This final hull form geometry is then fed back to the PAN system in order to carry out the final naval architectural calculations.

### 5.5.1.1 Description of the Tests

In a normal operating environment both PAN and SLAS operate on a PC under DOS, each with its own self contained database. Static interfaces exist to transfer the ship's hull form between these databases.

Within the testing environment the objective was to populate the NEUTRABAS database with the results of PAN and to replace the static "off-line" interfaces with interfaces through the NEUTRABAS architecture using as many of the modules as possible.

Two sets of interfaces were written, one for PAN and one for SLAS, each composed of a read and write module.

All of the tests were carried out in the DOS environment. However not all of the DMC modules were in place for this environment at the time of the tests. In order to overcome this, "ADS" (Application Development System) tools were used to generate modules which simulated the non-available DMC modules. These modules reproduced the functionality of the DMC by generating the equivalent read/write SQL commands for the ORACLE database.

Following the manual inputs of the ship's hull form data points into the PAN database through the PC keyboard, the sequence of stages in the test were as follows:

1. The UNIX based ORACLE database of the CADIS and CRESTA tests were reproduced under DOS. The same EXPRESS schema was loaded into the data dictionary.

2. The PAN application was executed which generated curves and calculated the naval architectural results which were then written to the PAN database.

3. The first phase of the interface read the PAN database and created an ASCII workfile.

4. ADS modules then read this workfile and instantiated the ORACLE database utilising the parallel express schema of the CADIS/CRESTA tests.

5. ADS modules again read from the ORACLE database and created a further ASCII file for subsequent input to the SLAS system.

6. The SLAS database was populated by an interface which read this ASCII file.

7. The SLAS application was invoked and the faired hull lines stored in the SLAS database.

8. The same process was then repeated to transfer the hull lines from the SLAS database to the PAN database, again using ADS modules to simulate the DMC functions.

### 5.5.1.2 Results and Observations

The original intention within the project was to develop NEUTRABAS tools primarily for the UNIX platform. However it was also hoped that similar tools could be developed to operate with MS DOS for PC applications. It was for this reason that the PAN and SLAS application interfaces were planned to work solely on a MS DOS platform. Unfortunately time did not permit the development of a complete set of MS DOS tools. Because of this it was found necessary to adopt the 4GL application development tool ADS to generate the equivalent NEUTRABAS access facilities and perform the equivalent functions upon the test schema entities.

Within these limitation the tests were successful in transferring all of the required information between the two systems. Indeed the use of the 4GL tool proved to be very successful and consideration could be given towards using such tools for future work of this nature.

Both interfaces communicated with the NEUTRABAS database through ASCII workfiles only because of the lack of DMC tools on the platform being used. No problems would have existed in dynamically connecting either of the two applications directly with NEUTRABAS had the tools been available.

The final observation concerned the elapse time to perform the bi-directional data transfer. The 1520 transactions generated during the tests took some 20 minutes on the AMD-486/33 PC used.

### 5.5.2  The CADIS and CRESTA Interfaces

The CRESTA system is a Project Control System which schedules parts through the manufacturing process using either the well known Precedence Diagram Method (PDM) or Arrow Diagram Method (ADM) techniques of representing the logical relationships between the activities which make up a complete or partial project.

The system provides information which enables management to ensure that a project is completed on schedule within budget, within the limits of certain constraints, such as the availability of resources (men, machines etc.), project dates and other budgetary considerations.

Typically the system can be used to address the following types of questions:

• If the only available resources are used, by how much will the project end be extended? (manpower constraint, resources limited).

• If the project end date is to be met, how many extra resources will be required? (time constraint).

• If overtime is worked on the project, by how much will the end date be brought forward? (A combination of time and manpower constraining).

• If activities dates are changed, what will be the effect on the project schedule and resources?

The system employs a two pass approach to the scheduling process, namely Time Scheduling and Resource Scheduling.

The Time Scheduling pass assumes unlimited resources and calculates the earliest and latest start and finish dates for each activity after the imposition of certain fixed dates on particular events or activities.

The Resource Scheduling pass assigns resources to each activity within the logical constraints of the network such that resource availabilities are exceeded by as little as possible.

The objectives of this two pass approach are:

• To determine the required amount of each resource at different time frames over a project's duration.

• To level the requirements for each resource to reduce peaks and troughs in the resource utilisation profiles.

• To calculate final start and finish activity dates.

Facilities exist within CRESTA to communicate with the other software systems, either by loading external files to its own master file or by creating transfer files for analysis by subsequent processes.

The CADIS System has a series of software modules designed to enhance productivity in the area of detailed steelwork design for ships structures. The system also addresses the creation of manufacturing information in an automatic or semi-automatic fashion for steelwork preparation, cutting, marking, assembly and erection.

In its normal working environment the CADIS process can commence as soon as a ship's hull form has been established. Ideally this should be a finished faired hull form like that produced by the SLAS application, but preliminary forms are sufficient to start the design process. Static interfaces exist with hull form definition and fairing software systems in order to populate the CADIS database.

Within a NEUTRABAS environment the geometry of the hull form would be retrieved from the NEUTRABAS database itself having previously been populated by such a system as SLAS. However timescales dictated the need to create the MOCK-UP model in CADIS early in the project before many of the NEUTRABAS tools were available and a direct interface to a flat file containing the geometry of the hull form was used. Following this, the steelwork model of the MOCK-UP was created using normal CADIS functions.

Working at the CADIS terminal the designer carries out a large proportion of his work using interactive graphics, the end result of which generates transactions which ultimately update the CADIS database. The transaction file is not permanent and is deleted after each update of the CADIS database.

Because the CADIS/NEUTRABAS interface was chosen to test the capabilities of NEUTRABAS in its role as a product data exchange medium, it was decided to use the CADIS database itself as the source of product information.

The NEUTRABAS (ORACLE) relational database, and all related modules and tools operate on UNIX workstations. In the case of the CADIS interface the chosen hardware platform was the IBM 6150 workstation using the AIX operating system.

Because two distinct hardware platforms were involved it was decided that the simplest and most practical method of operation should involve the use of intermediate files as illustrated in figure 1.

This decision brought with it a number of advantages during the implementation and testing phases as follows :

• Having established the format of the intermediate files, it was possible to develop interfaces I 1 and I 4 in parallel using two people not previously involved in the project. No in-depth knowledge of the NEUTRABAS architecture, software and tools was required other than a high-level appreciation of the EXPRESS schema. The intermediate file formats were deliberately made EXPRESSlike.

• Interfaces I 2 and I 3 (in fact a single module) were developed in parallel with I 1 and I 4 using someone familiar with the required DMC calls but without any knowledge of the application system. The resulting module is consequently schema independent and usable for other application interfaces.

This approach was considered as being more representative of a real-life situation than that of using someone who had lived with the project for some time.

• Early unit testing of the I 1 and I 4 interfaces could be carried out by simply going in and out of the intermediate file without going through the complete architecture of NEUTRABAS.

Within CADIS, additions, changes and deletions are handled by CADIS house-keeping facilities as the database is updated. The operation of the CADIS system does not lend itself naturally to testing the add/change/delete features of the NEUTRABAS system. In any case when testing NEUTRABAS's "product data exchange" capabilities it is considered that a "snap-shot" or "complete" data exchange is what would be required in practice. Because of this the main interface software concerned itself with transferring a complete CADIS model to and from NEUTRABAS.

However in order to simulate an on-line situation, where modifications are made to the original model, a special interface was defined which captured CADIS delete transactions, bypassed the CADIS database, and updated the NEUTRABAS database directly. These deletions would then reflect themselves back in the regenerated CADIS model.

The CADIS system, which begins at the early preliminary design stage and finishes with the detailed definition of all piece parts, (ie. top down) does not inherently hold the build sequence (ie. bottom up) within its database structure. In order to cater for this limitation and in particular to generate the NEUTRABAS "Prefabricated_Block" entities, a coding convention was adopted for those piece part entities (plates and commercial stiffeners) to be accessed subsequently by the CRESTA application.

Using the first 16 characters of the CADIS part number the following coding system was used for those piece parts which were part of a 'box' structure to be subsequently scheduled through the CRESTA application.

2090 DECK 0978 NNNN
2090 DECK 1238 NNNN
2090 LBHD 1095 NNNN
2090 LBHD 1365 NNNN
2090 TBHD 0215 NNNN
2090 TBHD 0219 NNNN
2090 TBHD 0223 NNNN

Where characters 1-4 were used to indicate the larger 3 dimensional prefabricated block to which the piece part belonged. In this case prefabricated block number 2090 which represented the complete box to be considered by CRESTA.

Characters 5-12 were used to indicate the smaller 2 dimensional prefabricated blocks which make up the 3D block. eg. DECK 1238 indicates the deck situated 12380mm above the base.

Characters 13-16 were used to uniquely define each piece part.

The result of this approach was that a NEUTRABAS model could be created which not only reflected the design with regard to its geometry, associated attributes and topology, but also reflected the sequence in which the product was to be manufactured and assembled. This approach was deliberately limited to a box structure within the mock-up so as to reduce the number of entity instances which were to be accessed by the CRESTA system. In this way the interactive nature of the CRESTA tests and demonstrations would benefit in terms of response times. The total number of entity instances in the mock-up was 10,300 whereas the box structure contained only 370.

### 5.5.2.1 Description of the Tests

The actual tests involved the three main computing platforms of mainframe, UNIX workstation and PC operating in a networked environment. The CADIS applications running on a remote IBM 308X running under MVS, communicated with the NEUTRABAS system operating on an IBM 6150 under AIX, via a 64KB kilostream line. The CRESTA application communicated with the 6150 through a local (NOVELL) network.

Following the creation of the mock-up model on the CADIS system the sequence of stages in the testing plan was designed as follows:

1. The installation of all the constituent modules of the NEUTRABAS architecture developed by WorkPackage 1 partners.

2. The importing of the EXPRESS Schema, defining the entities to be covered by the application interfaces, into the data dictionary.

3. The transfer of the test model from CADIS to NEUTRABAS.

4. The retrieval from NEUTRABAS of the test model and the creation of a new model on the CADIS system. Verification of the consistency between the original model and the newly generated model.

5. Retrieval from NEUTRABAS and the transfer of the relevant entities to the CRESTA scheduling system.

6. Running of the CRESTA application and the transfer of results to NEUTRABAS.

7.  Modification by the deletion of structural entities in the CADIS model and updating the NEUTRABAS database.

8.  Regeneration of the new model on the CADIS system. Verification of modified model.

9.  Re-running of the CRESTA scheduling process on the modified model. Comparison of results between original and updated schedule runs.

The CADIS to/from intermediate file interface software was written in FORTRAN. The intermediate file to/from NEUTRABAS interface software was written in C.

All of the CRESTA interface software was written in C.

Due to the limited length of time which was finally available to carry out the integration tasks it was not possible to carry out all of the tests as envisaged in the testing plan outlined above.

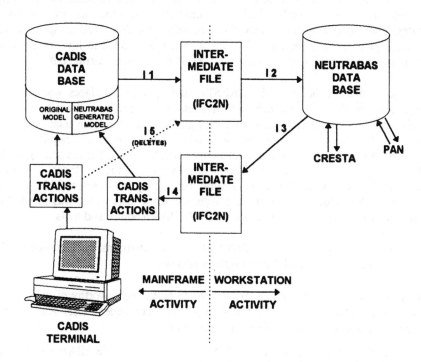

Fig. 65  CADIS Application Interface

## 5.5.2.2 Results and Observations

Nevertheless sufficient information in terms of entities, instances and attributes were transferred to fully validate the validity and stability of the NEUTRABAS architecture.

All of the constituent modules of the NEUTRABAS system developed by the Workpackage 1 partners were successfully installed on the UNIX hardware platform.

The data dictionary was then successfully loaded with the EXPRESS schema defining the entities included in the mock-up and to be referenced by the applications.

Unfortunately it did not prove possible to transfer the whole of the test piece, which had been modelled within the CADIS system, into the NEUTRABAS database. An examination of the intermediate file generated by the first phase of the CADIS interface showed that the whole mock-up stored in CADIS produced some 10,300 entity instances and 47,500 records which then needed to be processed by the DMC modules. The processing time involved on the relatively slow 6150 computer made such a complete transfer impracticable. As such it was decided to limit this part of the plan to a subset sufficient to enable the CRESTA tests to be carried out.

The major difference between the CRESTA and CADIS interfaces was that the former deals with fewer entities but invokes more functions. In addition the CRESTA application interface was interactive and as such more conversational in its nature. The limitation with regard to the number of entity instances populating the NEUTRABAS database referred to above only marginally affected the testing of the CRESTA application interface. Particular checks against the existence of certain entity instances had to be removed to allow processing to continue. This prudent approach in no way diminished the real objective of the tests which was to validate the functionality of the DMC and other NEUTRABAS modules and tools.

At the completion of the CRESTA tests both transfer programs worked successfully as did 6 update programs and 3 interrogation programs.

A number of specific observations or "lessons to be learned" worthy of mentioning arose during the development of the software and subsequent tests and demonstrations. The following are the most important of these:

• The two phase approach of the CADIS interface to NEUTRABAS enabled the second module to be independent of the application. Its role was to recognise such statements as CRE_SSR, ENTITY Geometry, END_ENTITY etc. and activate the appropriate DMC calls. This did mean however, that the logic associated with sequencing these instructions rested within the first module. This in fact proved to be one of the most troublesome parts of the implementation. The comparatively simple looking abstract schema has within it a number of sub-super relationships, and attributes whose values are themselves other entities and lists of other entities.

The resulting complex hierarchy of nested definitions needed to be treated with great care in order to ensure the accuracy of the fairly strict sequencing and syntax demanded by the calls to the DMC. Care should therefore be taken when defining schemas to limit the use of sub-super relationships and cross-referenced attributes to a minimum. A less demanding set of rules with regard to the sequencing of calls to the DMC would also assist when developing application interfaces.

• Another potential difficulty was highlighted during the development of the CADIS reverse interface (NEUTRABAS to CADIS). Unlike a forward interface, which knows precisely what information it has to pass and how it has to be structured, a reverse interface cannot be sure of what information exists within a NEUTRABAS database populated by a third party application. In the context of the CADIS tests this did not prove to be a problem as the system was effectively using NEUTRABAS as a means of exchanging product information with itself. In a true data exchange situation between two different CAD systems however, the logic of a reverse interface will need to be quite complex in order to cover all possible eventualities.

• The final and most important observation to be made concerns the use of NEUTRABAS as a medium for data exchange. Based on the number of entity instances generated by the mock-up, it is envisaged that the number of entity instances for a complete ship's structure could be many millions. The conclusion to be drawn from this is that the processing time involved, even when using the most powerful processors, could be prohibitive if a complete ship's structure were to be exchanged. This conclusion will need to be verified by more comprehensive testing.

# 6 Experiences and Results

## 6.1 Overview

At the time when the Technical Annex and Work Programme for NEUTRABAS were written in 1989 the formal information modelling language EXPRESS and the forthcoming international standard STEP (ISO 10303: Industrial Automation Systems - Product Data Representation and Exchange) were both still undergoing significant change and were far from stable workying platforms. It thus had to be anticipated by the NEUTRABAS project team that many of its requirements in developing a neutral database for shipbuilding applications could not be met initially by the standard as it then existed. It was therefore regarded necessary to document those requirements of NEUTRABAS which went beyond the current capabilities of EXPRESS and STEP, especially if they still were not met by the end of the NEUTRABAS project. This is the original objective of Subtask 1.2.7 in NEUTRABAS: "Required Extensions to EXPRESS and STEP", concentrating on the required extensions needed to support a neutral database and reference model for shipbuilding.

When NEUTRABAS ended in 1992 EXPRESS (ISO 10303-11, Part 11: Descriptive Methods: The Express Language Reference Manual, Document N15 of ISO TC184/SC4/WG5, 29 April 1991) had been approved as Draft International Standard and was essentially stable. A few other parts of STEP had also already achieved DIS status (Part 21: Clear Text Encoding, Part 42: Geometric and Topological Representation), most other Parts intended for the initial release of STEP reached DIS status in 1993. This means that the12 initial parts of STEP, including the essential parts for description and implementation methods and generic resources, will become International Standards (IS) in 1994. The STEP Standard thus is much more advanced than in 1989, though clearly many capabilities are still missing.

NEUTRABAS has been developed during this period of frequent change and has had to adjust itself to many updates. Many of the requirements which the earlier drafts of the standard did not meet have been taken into account in the meantime. NEUTRABAS participated in many ISO discussions and was able to express its priority needs. In many cases these needs were similar to those held by other industries and product data user communities. Thus the priorities in user requirements and the current shortcomings in product modelling based on EXPRESS and STEP have become much better understood by the STEP community. Some relief actions have been taken, other open questions and problems persist.

Due to these developments the main focus in this subtask of NEUTRABAS has had to shift. The report given in this chapter reflects the status reached in 1992 and has the following main objectives:

• Relate the status of EXPRESS and STEP and the required extensions which are proposed in the STEP community and are also needed by NEUTRABAS.

- Give an account of specific NEUTRABAS experiences with shortcomings of EXPRESS and STEP.

- Develop recommendations for future STEP extensions to meet specific needs in shipbuilding.

The chapter, however, does not intend to give a comprehensive evaluation of EXPRESS and STEP from a general viewpoint that goes beyond the scope of NEUTRABAS.

## 6.2 Extensions to EXPRESS

### 6.2.1 Status

EXPRESS, as defined in ISO DIS 10303-11 [23], is a conceptual schema definition language. It can be used in particular to describe information models of product model data. Within certain restrictions it can also be regarded as an object-oriented data definition language (OODDL), although it does not support the full object-oriented paradigm as will be discussed below.

During its development EXPRESS has been subject to numerous and frequent changes so that it was difficult to be used as a definitive modelling tool. Since its DIS approval in 1991 it has been essentially stable, it has still somewhat further modified in some minor matters before reaching IS level but its functionality did not substantially change any more for the initial release of STEP. This is of great importance to secure the investments in information models using STEP.

Despite this stability of the language which is desired for many practical reasons it is already well recognized that EXPRESS does not meet all product modelling requirements which exist today and, in fact, that its limitations have been an impediment to several more advanced applications. Certain information models which are perfectly legitimate from an application viewpoint cannot be fully described using EXPRESS in its current form. Nevertheless it has been decided by ISO to defer any action toward further extensions of EXPRESS until the standard STEP has reached maturity and acceptance with the many models that can already be supported by EXPRESS as it is.

Simultaneously, however, ISO has started an initiative in April 1991 to begin the development of EXPRESS Version 2 by systematically gathering user requirements for future extensions of the language. This work is performed under the guidance of WG5 (STEP Development Methods) of ISO TC184/SC4. It happens that two representatives of the NEUTRABAS project team, Bernd G. Wenzel and Dr. Sabine Müllenbach, are members of this task group so that communications between this activity and NEUTRABAS were excellent in both directions. After three major international workshops (Munich, Sapporo, Charleston) and ISO level discussions in Houston and Oslo a Working Draft Document has been issued for this group by

Wenzel and Müllenbach. It is entitled "EXPRESS Version 2 - Requirements and Project Proposal", version 0.4 [24].

This document gives a comprehensive overview of current user requirements for extensions to EXPRESS. It will be summarized and discussed in section 6.2.2. In practice all NEUTRABAS requirements for EXPRESS are included in this catalog. Only the priority and preferred solution may sometimes deviate. Therefore the discussion of proposed extensions by NEUTRABAS can follow the basic approach taken for EXPRESS Version 2.

It should be realized, though, that the project plan for EXPRESS Version 2 requires resources of 18 manyears and a time frame of about 6 years. For NEUTRABAS many of the shortcomings addressed here are of immediate concern. Therefore one must also call for interim solutions and perhaps an accelerated schedule for Version 2 of EXPRESS.

### 6.2.2 EXPRESS Version 2

In a series of workshops held by WG5 of ISO TC184/SC4 ten important requirements and proposed solutions for extensions of EXPRESS were collected and documented [24]. The following list will give a brief summary with comments from the viewpoint of NEUTRABAS:

Requirements for EXPRESS Version 2

1. *Type Declarations*
   Type declarations have to be made more flexible in some respects, more stringent in others. For example, modeler defined types should be clearly distinguished from base types. Type specific operations shall be supported. Operation inheritance from representation types must be controlled etc.

2. *Type System Extension*
   Typed expressions, typed functions and maybe typed procedures should be treated as dynamic data items so that they can be activated by some operator function in a receiving system. This is an important requirement since it opens the way for the language into procedural applications such as parametric design. It remains open whether this feature can be offered in a recursive form.

3. *Operations and Methods*
   EXPRESS needs operation definitions, probably both for binding within entities where the operations can be invoked like a method and independently for defining methods in some modular form.

   This requirement will serve to support more fully the message paradigm of object-oriented languages.

4. *Runtime Behaviour Specification*
   To be able to define the runtime behaviour of an EXPRESS implementation functions to describe the database behaviour in response to basic operations and more complex database manipulations are needed.

5. *System Modelling*
   This important requirement refers to a major extension of EXPRESS from a static information modelling language to a dynamic systems modelling language. This extension will have to account for such concepts as:

   Action, event, time, trigger, message, sequence, concurrency.

   This is an important prerequisite for the modelling of time-dependent functional systems.

6. *Schema Transformation Language*
   The ability to map EXPRESS schemas onto one another is currently limited to identity entity mappings in the language. But complex mappings of schemas are often needed, e. g., to describe the ARM to AIM mappings or to map concrete application system schemas onto STEP conceptual models. A formal language for describing these transformations would be very useful.

7. *Increase of Meta Levels*
   By generalizing the two-level distinction between types in a schema and instances in a database in such a way that types can become instances of higher level types one obtains a powerful, yet abstract and ambitious extension of EXPRESS.

8. *Descriptive Constraint Language*
   The language should be made more descriptive with regard to constraint formulation with the objective of stating pre- and postconditions for the state of the database before and after execution of a procedure. This may reduce the need to mix procedures into the information model.

9. *Implementation Hints*
   To convey advice by the modeller for good or precise implementations of the model. Doubts exist about the usefulness and possible rigor.

10. *Normative Meta Model*
    The existence of a normative EXPRESS meta model would be useful in standardizing the language and its growth.

These ten recommended extensions in EXPRESS Version 2 cover a large scope and would increase the capabilities of the language by a significant factor. However, they are not necessarily all compatible with each other and are not easy to implement and support. Therefore it is necessary to set priorities and to aim at a realistic scope of extensions in the near future. Evidently not all of the proposed extensions which it would be nice having are necessary for today's needs in product data modelling.

Therefore from the viewpoint of NEUTRABAS and its experiences the next section will evaluate important needs and set priorities in this spectrum of possible extensions.

### 6.2.3 NEUTRABAS Experiences and Recommendations

During the NEUTRABAS project comprehensive experience was collected in the use of EXPRESS for the development of information models and hence databases. EXPRESS was still under development itself and had many weaknesses and insufficiencies in the beginning. But it matured steadily and was successfully used as the main information modelling tool in NEUTRABAS. A summary of the experiences gained during these developments is given in the following. This includes in particular the results already documented in Deliverable 3.2.3 [25] (Synthesis of the Specification Activity).

Based on these experiences the following limitations in the current functionality of EXPRESS have to be recognized and, if possible, overcome in future EXPRESS versions:

#### 6.2.3.1 Scope of the Language

When comparing available information modelling languages and methods three categories can be distinguished:
a) Semantically Irreducible Information Modelling:
   E. g.: NIAM, INFODIC, Entity Relationship Models
b) Record Oriented Information Modelling:
   E. g.: IDEF1X, Relational Models
c) Object-Oriented Information Modelling:
   E. g.: EXPRESS

EXPRESS permits the description of a static information model, i. e., a conceptual schema for a database, and has a strongly object-oriented flavour. But it must be noted that it does not support the full paradigm of object-oriented languages. These languages are characterized by the following capabilities:

1) Definition of objects (objects carrying information as attributes, also called entities or abstract types)
2) Inheritance of attributes (simple or multiple, from supertypes to subtypes)
3) Embedding of rules (procedures, methods) in objects permitting the object to respond dynamically to a message so that its state may depend on the execution of procedures
4) Exchange of messages between objects as a means of updating the information base

EXPRESS does possess the former two, but not the latter two capabilities, which was already noted as a limitation in Deliverable 3.2.3.

Further it is also understood that EXPRESS is not a procedural programming language because it lacks functions for input/output, data value assignment, execution error handling etc. These functions are, of course, deliberately omitted.

EXPRESS is thus a suitable language for the description of static (passive) information models by conceptual schemas, e. g., for data exchange purposes. It has currently serious limitations with regard to capturing dynamic, changing information models.

It is interesting to observe that Version 2 of EXPRESS addresses precisely these limitations and seeks to remove them. Items 2, 3 and 5 in its requirements list are all designed to turn EXPRESS into a more full-fledged object-oriented language that can be used for modelling systems with time-dependent and procedural information.

Even if all of the language extensions currently contemplated for EXPRESS Version 2 are eventually realized, there remains a general concern shared by many NEUTRABAS partners who have used EXPRESS as a modelling tool: There exists a large gap already between a subject expert's conceptual understanding of a product information structure and the abstract, formal and rigourous view which the modelling language requires. This gap may be widening when more and more functional capabilities are added to EXPRESS. Even now somebody who understand perfectly well what a trimmed rational B-spline surface means does not recognize it any more when he encounters a representation item whith a representation relationship etc. Thus the subject expert becomes divorced from the modelling process.

This raises the issue of whether the language can be extended or even reduced to a metalevel in such a way that the modelling process would be supported in a more intuitive way and required less rigour at model design time. The precision of the model might be augmented later by partly automatic tools.

Finally it must be postulated to keep Version 2 of EXPRESS upward compatible to Version 1 to the greatest possible extent. Otherwise the substantial investments in EXPRESS models of the current generation would be threatened or lost. Therefore a policy must be established before embarking on changes for Version 2 as to which features of Version 1 must be retained. Current users of EXPRESS should be warned now if any features of the language are likely to be removed later.

### 6.2.3.2 Technical Difficulties

Many original criticisms of EXPRESS by NEUTRABAS partners stem from technical difficulties encountered in its practical use. This was mainly due to the lack of a stable language definition and reliable EXPRESS tools. A variety of EXPRESS parsers and compilers used at different stages of the project (mainly CADDETC LUPINE, McDONNELL DOUGLAS, NIST FED X). They refer to different language releases of EXPRESS, usually not the latest document, and have different omissions and levels of reliability. Usually a schema translation by different parsers will produce different error lists. These difficulties still exist, no complete and up to date EXPRESS tools seem to exist yet, although improvements can be expected now that EXPRESS is stabilized as DIS.

It was also frequently criticized that no standard tools exist for EXPRESS based database implementations. A data manipulation language (DML) and a data control language (DCL) as extensions of EXPRESS would be very useful to create data base management systems for EXPRESS models. Without these tools each implementation is a much more tedious manual programming effort. Version 2, requirement 4 addresses this need.

Much time in NEUTRABAS was also devoted to developing mappings from concrete schemas existing in communicating CAD systems and the conceptual schema of the neutral database. A general schema translation language as proposed by requirement 6 of Version 2 would have saved much effort. These mappings will occur at least once for each newly connected CAD system and will therefore require much repetitive effort.

### 6.2.3.3 Modelling Difficulties

NEUTRABAS also encountered some fundamental modelling difficulties which were caused in part by limitations in the EXPRESS language, in part also by lack of suitable STEP resources. The most important NEUTRABAS modelling requirements which cannot be fully met by current EXPRESS and STEP methodology are the following:

- Access to libraries
- Parametric design parts
- Product functionality

All three of these capabilities depend on extensions in EXPRESS considered for Version 2, especially the ability to define operations, procedures, messages and other dynamic functions. Provisionally in NEUTRABAS as in other STEP projects only a static solution could be offered for these purposes. This cannot be accepted as adequate.

The matter will be discussed in more detail in the following section.

## 6.3  Extensions to STEP

### 6.3.1  Status

The initial release of the DIS for STEP (ISO 10303) is scheduled to complete DIS balloting by December, 1992. The initial release will include the following 12 parts:

Part No.   Title

1          Overview and Fundamental Priciples

Description Methods:
11         The EXPRESS Language Reference Manual

Implementation Methods:
21         Clear Text Encoding of the Exchange Structure

Conformance Testing:
31         Conformance Testing Methodology

Integrated Generic Resources:
41         Fundamentals of Product Description and Support
42         Geometric and Topological Representation
43         Representation Specialization
44         Product Structure Configuration
46         Visual Presentation

Integrated Application Resources:
101        Draughting

Application Protocols:
201        Explicit Draughting
203        Configuration Controlled Design

These parts will secure a basic foundation for all further, more application-oriented developments in STEP. Many additional parts of STEP are already far advanced and will come to DIS level maturity soon after the initial release. This next generation of STEP Parts will also include Application Protocols in Design for BRep, Surface and Wireframe Representation, hence the exchange standards for the great majority of present CAD systems in mechanical design. A complete list of proposed STEP Parts is attached as Table 2.

Regarding developments in STEP for the shipbuilding and other maritime industry sectors only one part, viz. Part 101: Ship Structures, which was submitted by NIDDESC, belongs to the present set of proposed STEP Parts. However, at the ISO Meeting in Oslo in February 1992 six additional Application Protocol Planning Projects, i. e., preparatory developments toward an AP, were submitted by NIDDESC and accepted by PMAG. They are:

Ships Electrical Systems
Ships Heating, Ventilation, Air Conditioning Design
Ships Library Parts
Ship Outfit and Furnishings
Ship 3D Piping
Ship Structural Systems

These proposed Application Protocols intend to make use of the Integrated Generic Resource of Part 102, Ship Structures, where applicable, so that there should be no conflict between the ship structures schemas at resource and at AP Levels.

**Table 2**   STEP Parts Status - December 1991

| Part | Title | Initial Release Next CD Date | Proposed Next CD Date |
|------|-------|------|------|
| 1 | Overview & Fundamental Principles | 01/92 | |
| | *Description Methods:* | | |
| 11 | The EXPRESS Language Reference Manual | DIS | |
| | *Implementation Methods:* | | |
| 21 | Clear text Encoding of the Exchange Structure | DIS | |
| 22 | STEP Data Access Interface | | |
| | *Conformance Testing:* | | |
| 31 | Conformance Testing Methodology and Framework (CTMF) - General Concepts | 11/91 | |
| 32 | CTMF - Requirement on Testing Laboratories and Clients for the Conformance Assessment Process | | Apr. 93 |
| 33 | CTMF - Abstract Test Suite Specification | | Jul. 92 |
| 34 | CTMF - Abstract Test Methods | | Apr. 93 |
| | *Integrated Generic Resources:* | | |
| 41 | Fundamentals of Product Description and Support | CD | |
| 42 | Geometric and Topological Representation | DIS | |
| 43 | Representation Specialization | CD | |
| 44 | Product Structure Configuration | CD | |
| 45 | Materials | | Oct. 92 |
| 46 | Visual Presentation | CD | |
| 47 | Shape Tolerances | | Jan. 93 |
| 48 | Form Features | | Jan. 93 |
| 49 | Product Life Cycle Support | | |
| | *Integrated Application Resources* | | |
| 101 | Draughting | CD | |
| 102 | Ship Structures | | Apr. 93 |
| 103 | Electrical Functional | | |
| 104 | Finite Element Analysis | | Oct. 92 |
| 105 | Kinematics | | Oct. 92 |
| | *Application Protocol:* | | |
| 201 | Explicit Draughting | 12/91 | |
| 202 | Associative Draughting | | Oct. 92 |
| 203 | Configuration Controled Design | CD | |
| 204 | Mech. Design Using Boundary Representation | | Jan. 93 |
| 205 | Mech. Design Using Surface Representation | | Jan. 93 |
| 206 | Mech. Design Using Wireframe | | |
| 207 | Sheet Metal Dies/Blocks | | |
| 208 | Life Cycle Product Changes Process | | |

Key: CD:   Committee Draft - part is currently in international review
   DIS:   Draft International Standard - CD review  if part resulted in concensus approval
   xx/xx:  Planned date or next CD review start

Note: All Parts in Initial Release column must reach DIS before any proceed to International Standard.

Evidently the spectrum of proposed APs for shipbuilding is in almost complete overlap with the scope of the NEUTRABAS project. While in the past NIDDESC used to concentrate primarily on the detail design and production stages in the life cycle of these applications, the proposed new scope of the APs now comprises all life cycle stages from functional design to support engineering. Naval shipbuilding practice forms a strong basis and motivation. All proposed part owners are NIDDESC representatives.

This situation must be reviewed by the NEUTRABAS team both technically and procedurally. Technically it is a positive development that ISO has accepted these shipbuilding application areas as AP standardization subjects. The list of subjects had been agreed upon between NEUTRABAS and NIDDESC earlier as meriting AP developments.

However, there are still wide gaps in positions regarding the organisational structure, allocation of schemas and STEP Parts, and certainly the technical substance of each part. NEUTRABAS takes the position that these conflicting views cannot be reconciled and a meaningful division of labour and responsibility cannot be obtained without first agreeing on a general, but detailed enough reference model for all shipbuilding applications. NEUTRABAS has developed a framework draft for such a model in its Deliverable 1.2.6, whose principal concepts will be explained in the following section.

### 6.3.2  Required Application Schemas for Shipbuilding

### 6.3.2.1  A Proposed Architecture

NEUTRABAS and NIDDESC have both taken the approach of developing rather independent information models as nearly self-contained schemas for each separate application area, which are then candidates for Application Protocols. This is essentially a bottom-up approach to information modelling. In the experience of NEUTRABAS, but also from observation of the NIDDESC document set, this approach does not result in a coherent perspective of all APs and in an integrated, interoperable set of application schemas. NEUTRABAS therefore has adopted the position that an overall schema architecture for the shipbuilding application area with a comprehensive ship application reference schema is needed as a framework. This reference schema provides a top-down perspective of this application area and is required to succeed in integration.

The fact that many application schemas share the same generic STEP resources is not sufficient to ensure consistent integration. In practice many application schemas use the same resources in very heterogeneous ways. They also create their own sets of application oriented resources which tend to become redundant and contradictory. The desired order in this havoc of application schemas can only result from a top level perspective of a schema architecture where responsibilities are clearly assigned at the subschema level.

As a result of these considerations NEUTRABAS is proposing a general schema architecture for shipbuilding as shown in Fig. 2. The architecture is composed of four levels:

• Ship Application Reference Schema
• Specific Ship Application Schemas
• Ship Application Resource Schemas
• STEP Integrated Generic Resources

In terms of conventional STEP architecture these four levels have the following correspondences:

Top three levels: One or several ARMs
Bottom level: Used for AIMs
All four levels: One or several APs

In this way an Application Protocol can be built up from numerous schemas as building blocks. In fact, the modules shown in Fig. 2 may be subdivided further to arrive at the appropriate granularity of schemas within this architecture.

More elaboration on this schema architecture, especially at the levels of the Ship Application Reference Schema and the Specific Ship Application Schemas, is presented in Deliverable 1.2.6 of NEUTRABAS.

The four levels in this architecture have the following functions:

• **Ship Application Reference Schema**
  This schema is a conceptual schema written in EXPRESS that provides the basic organisation for ship product definitions. It contains not only high-level abstract concepts which are needed to subdivide the tasks of ship product description, but also a framework of specific entities which serve as integration points between the organisational reference schema and the specific ship application schemas. NEUTRABAS Deliverable 1.2.6 provides the first draft for such a reference schema.

• **Specific Ship Application Schemas**
  This level contains schemas for specific application areas. Each area may be covered by one or several subschemas with suitable integration points. This schema level is connected with the reference schema below by entities serving as integration points. The schemas are defined in the style of ARMs.

• **Ship Application Resource Schemas**
  These schemas comprise information that is shared by several application schemas. Ship geometry and topology is in common to almost every application, so it is the most likely candidate for an integrated application resource. But by close analogy many other common resources can be found for ship applications, in particular those which have their equivalent schemas at the level of STEP Generic Resources. In this way the shipbuilding perspectives of all resources would be unified at the

application resource level, which could then be directly mapped down to STEP generic resources.

- **STEP Integrated Generic Resources**
For this level a suitable subset of the available resources from Parts 41, 42, 43, 44, 45, 46, 47, 48 and 49 must be selected and described in AIM fashion to serve as a pool of entities for STEP Application Protocols in shipbuilding.

More details about possible assignment of schemas and entities to the various levels in this architecture can be found in NEUTRABAS Deliverable 1.2.6 where a framework for the realization of this schema architecture is presented.

### 6.3.2.2 Proposed Application Protocols and Resource Schemas

Within the framework of general schema architecture for shipbuilding and under the common roof of a ship application reference schema the following *Application Protocols* should be realized in STEP:

For global ship information:
- General Ship Chaacteristics

For physical product definition:
- Spatial Arrangements
  (possibly subdivided and with auxiliary schemas

- Ship Structural Systems

- Outfitting Systems
  (with subschemas or independent APs for machinery, hull engineering, furnishings, HVAC, piping, electrical and other shipboard systems)

For functional product definition:
- Ship Functional Systems
  (probably with many subcategories)

For product activity definition:
- Shipbuilding Activities
  (such as design, production engineering, production, support etc.)

These various APs will form a coherent set by sharing both a reference schema and integrated application resources.

The following schemas are candidates for becoming *Ship Application Resource Schemas* (at the level of 100 Parts in STEP):

- Ship Geometry and Topology
  (for volume, surface, and wireframe models)
- Ship Product Structure Configuration
- Ship Materials
- Ship Product Life Cycle
- Ship Standard Libraries

It is conceivable to form one or several Application Interpreted Constructs (AICs) from these pools of entities. The status of AICs as integration devices in STEP and their modelling in EXPRESS have not been fully clarified yet.

### 6.3.3  Required Generic Resources

The schemas involved in shipbuilding APs will be based on the available Integrated Generic Resources in STEP which are defined in Parts 41 to 49 of STEP and are gradually approaching DIS status. Shipbuilding will require a suitable subset from this pool.

In addition, by extensions of STEP further generic resources must be made available to cover all requirements for the shipbuilding APs. The main areas where it is already apparent now that further resources are needed than STEP can provide now are the following:

– Access to libraries
– Parametric design and standard parts
– Product functionality

These requirements are so general in STEP that a solution at the generic level must be provided. Appropriate functions should be included in STEP's Integrated Generic Resources although extensions to the EXPRESS language are also required for many of these purposes as was discussed in section 2.

Let us examine these resource requirements in some more detail.

### 6.3.3.1  Library Access

In order for a product data exchange in STEP to be supported by libraries available at the receiving end (and known to the sender) the following conditions must be met:

* An *external reference* must be established to some library, which is not part of the data exchange, but exists in the operating environment of the recipient. A library name must be understood by the receiving system.

* An *addressing scheme* for a library item must be uniquely identified. If the library has a complex structure, the addressing scheme must be of according structure.

* The *insertion of the library* item into some context in the STEP data exchange requires a transformation at the instance level which maps the exchange file parameters for the item onto the instance occurence level of the item. This infers that a procedural step is executed by the receiving system which cannot be described in the EXPRESS language, but must still be uniquely understood by the participants in the data exchange. An example is the display of a graphical symbol by a receiving graphics system where the data sent are sufficient to define the desired appearance of the symbol without having to state the algorithm for its display.

These three conditions are the minimum requirements for the use of libraries in STEP. They will be sufficient only if procedures to invoke the libraries need not be explicitly stated because they are known. This will be called a static, purely data based library access. (Dynamic, procedurally based library references are dealt with under the next heading).

Currently STEP provides none of the three required mechanisms in a suitable form. The responsibility for part libraries and library structures lies with Working Group 2 of SC4. They have produced useful documents on objectives for part libraries [26] and library structures [27], but no coherent, EXPRESS based design for a solution. Clearly they suffer from limitations in EXPRESS for their purposes, too.

In WG2 consensus was reached at least that the conceptual model for standard part libraries developed by the European CEN/CENELEC CAD-LIB project is the most advanced, "technically complete" concept on which future developments will be based. This approach, too, will require extensions to EXPRESS and STEP Physical File.

Returning to the subject of static library access there exists now a method for external references based on STEP Physical File (Part 21), where an external reference by computer is supported. This will permit to exchange a library name as a string of characters. In the example of a text font library reference the following entity could be defined by those means:

```
ENTITY font_lib_ref
   library_name     : external_reference_by_computer:
   font_family      : identifier;
   font_name        : identifier;
   alphabet_name    : identifier;
   string           : string;
END_ENTITY;
```

This method would meet the three conditions for static library access, but was rejected by STEP Integration and PMAG in Seattle. An ad hoc group was appointed instead to develop an approach called Data Transfer Model by which libraries can be accessed. The results are due to be reported to ISO in London in July. They are urgently awaited because the Initial Release of STEP needs libraries for several of its parts.

### 6.3.3.2 Parametric Design and Standard Parts

Parametric design capabilities are offered by many CAD systems, but are not supported by STEP yet. The procedural functions which are needed to define the algorithms by which the parametric object can be made explicit cannot be described in EXPRESS. For NEUTRABAS this ability would be very relevant for parametric design of ship systems, parts and assemblies.

Standard part libraries which are a frequent occurrence in ship design, particularly for structural elements, can be regarded as a special case of parametric design. The parameters by which the library item is identified also serve as inputs to the algorithm which explicitly defines the parametric part. This evaluation corresponds to a dynamic library reference.

This type of library reference for standard parts is an essential requirement for NEUTRABAS. The objectives are similar to those developed by the CAD-LIB project of CEN/CENELEC. Suitable STEP and EXPRESS extensions are urgently needed for this purpose.

### 6.3.3.3 Product Funtionality

The way the product functions can be viewed today in STEP is limited to a static data model of product functionality. This approach had to taken for certain modelling tasks in NEUTRABAS. But it was recognized, too, that modelling of product functionality in terms of systems behaviour and processes was not achievable within the current scope of EXPRESS. In other words, the lack of modelling tools for operations, events, messages, triggers, time-dependence etc. prevents anything better than a flat, static view of product functionality. Or in a slightly more positive spirit, the static data view of product functions describes product functionality only to the extent that product functional behaviour can be uniquely inferred from the data.

Within the organization of ISO TC 184 SC4 it is a special interest group for Product Functionality in WG3 that has assumed responsibility for this subject in STEP. A STEP project may be in preparation. This group has provided a few principal definitions:

- Functionality: The set of all functions defined for an arbitrary instance of a product.

- Function: A rule relating an initial product condition to a resulting product condition triggered by some event.

If this paradigm or a similar one is adopted to model functionality then evidently a modelling tool is required which has the ability to describe system states, events, and time dependence, in short, processes.

For future functional models of shipbuilding processes NEUTRABAS therefore endorses extensions to STEP and EXPRESS that provide a generic capability for modelling functional systems and processes.

### 6.3.4 Experiences with STEP Integration

The experiences with STEP Integration within NEUTRABAS stem from two sources: NEUTRABAS' own activities in coordinating and integrating its evolving applications schemas and secondly, observations by some of its members of the ISO integration process as it affects the initial release of STEP. The following comments refer to both of these experiences.

Many of the developments under STEP were based on the premise that it is feasible to build a house by first building the individual rooms separately by independent teams and then calling an architect who will put the house together. Even the fact that everybody must use the same bricks as resources does not help the integrating architect very much.

The danger which lies in this approach was underestimated by the STEP community when it relied too exclusively on a bottom-up approach for which the object-oriented flavour of EXPRESS as an information modelling tool was only too tempting. As a result a heavy penalty had to be paid later when integration came in at a stage when the committee drafts of the major STEP Parts were ready.

NEUTRABAS went through very similar stages of development and on its smaller scale had similar experiences. It turned out to be extremely difficult to integrate the independent application schemas afterwards when a reference schema was to be developed in the end. This difficulty was only somewhat alleviated by the fact that the application experts and the data modelling specialists worked closely together and were sometimes the same people.

The obvious lesson from this experience should be that an application reference model must be developed for any major application area from the very beginning. It may have to be modified as the work progresses, but it will provide top-down guidance to the bottom-up working teams.

But aside of this main procedural criticism of the STEP integration process, there were also many technical and practical difficulties which have resulted in much delay. This report will name only a few of the main impediments encountered with STEP Integration. From our experience these are:

### 6.3.4.1 Lack of Clear Methodology

There exist no written statements with clearly and completely delineated objectives, criteria and guidelines for STEP integration from which a methodical approach to integration can be developed. Consequently neither the document owner nor a neutral outsider can judge when integration is successfully completed.

### 6.3.4.2 Deviation from EXPRESS

STEP Integration establishes internal, often unwritten rules how EXPRESS should be used. This narrows down the scope of the language which was approved as STEP DIS. At least such usage style guidelines should be proposed in writing and approved at STEP policy levels.

### 6.3.4.3 Preponderance of Geometry

Within the generic resources of STEP geometry/topology were among the first developed and are certainly playing a central role in product modelling. This poses no difficulties.

However, STEP Integration is still making changes to these central parts (Parts 41, 42, 43) at this very late stage which will have to propagate into many other STEP Parts later.

Moreover, many of the constructs and entities of these parts, notably 41 and 43, are used by Integration in the manner of templates into which other parts must be molded.

This forces other parts into complying with a certain style of modelling which was not declared in advance, but was imposed retroactively as Parts 41 and 43 gradually developed. Recent revisions to these documents and their interpretation aggravate this problem.

### 6.3.4.4 Atomistic Modelling

In object-oriented information modelling it is legitimate to introduce objects (entities) which are not the smallest decomposable units of information, but reprensent some higher semantic level.

In STEP Integration there seems to exist a tendency to favour a modelling style where the information is broken down into its atoms. This results in an unnecessary proliferation of entities which reduces the clarity of most models from an overall viewpoint.

Even where such modelling choices are not just a matter of style preference, there should be clear policy statements issued when and why an entity needs to be broken up.

### 6.3.4.5 Missing Documentation of STEP Approach

STEP lacks a clear and complete description of its methodology which explains how a new standard information model is developed. Precise definitions for the application protocols and their subsets, relationships between different models, between ARMs and AIMs etc., an overview of existing generic resource entities and their use would all be required to facilitate the use and developments of the STEP standard. These needs have been recognized by the STEP community for some time and presumably will be met soon as STEP becomes a stable and mature standard.

## 6.4 Conclusions

During the three years of NEUTRABAS development EXPRESS and STEP have undergone significant changes, which for the most part were improvements and necessary corrections. NEUTRABAS has had to go through several cycles of update, but was able to adapt to changing conditions and methodologies in the standard. But even today there still exist many shortcomings and open questions in ISO 10303 so that the requirements of NEUTRABAS and the shipbuilding community cannot be fully covered. This report summarizes the experiences gained by NEUTRABAS and the recommendations given for future extensions to EXPRESS and STEP. The main conclusions and recommendations are as follows:

EXPRESS:

• NEUTRABAS endorses the requirements for EXPRESS Version 2 as given by WG5 of ISO TC 184/SC4. This includes changes to the language for modelling operations, methods, and procedures, for system modelling, schema transformation by formal mapping and a number of practical improvements.

• NEUTRABAS has experienced several difficulties due to lack of a stable standard, well-defined resources, and effective tools based on EXPRESS. This should include standard technologies and interfaces to create database administration systems for STEP.

• The language EXPRESS should be extended in particular in such a way that access to libraries, parametric design, standard part definition, and modelling of functional properties of systems can be readily realized.

STEP:

• NEUTRABAS proposes a general schema architecture for shipbuilding applications, based on the experiences described in its Deliverabe 1.2.6, which consists of four levels:

    – Ship Application Reference Schema
    – Specific Ship Application Schemas
    – Ship Application Resource Schemas
    – STEP Integrated Generic Resources

This architecture should be adhered to in organizing all shipbuilding applications in the standard and as a framework for coordinating contributions by NEUTRABAS, NIDDESC, and others.

• The set of Application Protocols proposed on this basis includes the following six APs:

    – General Ship Characteristics
    – Spatial Arrangements
    – Ship Structural Systems

  - Outfitting Systems
  - Ship Functional Systems
  - Shipbuilding Activities

In addition the following five Ship Application Resource Schemas, which are shared by several ship applications, need to be developed:

  - Ship Geometry and Topology
  - Ship Product Structure Configuration
  - Ship Materials
  - Ship Product Life Cycle
  - Ship Standard Libraries

• STEP must be extended also with regard to generic resources to include capabilities for

    Library access
    Parametric design
    Standard parts
    Product functionality

• The integration process in STEP must be restructured in such a way that transparent entities and guidelines exist so that developers can verify themselves whether they meet integration requirements.

• The use of reference models will serve as an important instrument for integration.

• NEUTRABAS as a team of partners must make plans and set goals for the period following the expiration of the contract in order to exploit its own results and promote the positions in standardization for shipbuilding which it has developed in the framework of STEP.

# Literature

1. Gerardi, M. L.: Application Reference Model for Ship Structural Systems. Working Draft of Part 102 of ISO 10303 (STEP), ISO TC 184/SC4/WG1, Version 4.0, June 1990.

2. Martin, D. J.: Reference Model for Distribution Systems, Version 1.0. Working Draft, ISO TC 184/SC4/WG1, Doc. N443, December 1989.

3. Schenck, D.: EXPRESS Language Reference Manual, Version 1990. Working Draft, Doc. N496, ISO TC 184/SC4/WG1, 1990.

4. Nijssen, G. M.; Halpin, T. A.: Conceptual Schema and Relational Database Design, A Fact Oriented Approach. Prentice Hall, 1989.

5. Palmer, M.: Guidelines for the Development and Approach of STEP Application Protocols, Version 1.0. ISO TC 184/SC4/WG1, Doc N34 (P5), Draft, 20 February 1992.

6. Brèche, P.; Brun, P.; Cazaux, A.; Lehne, M.; Lynch, A.; Nowacki, H.; Vopel, R.: Shipbuilding Reference Model Framework. NEUTRABAS Del. 1.2.6, ESPRIT Proj. No. 2010, July 1992.

7. Taggart, R. (ed.); Ship Design and Construction, SNAME, 1980, New York (USA).

8. NEUTRABAS Deliverables 3.1.1/2/3, 3.2.1/2.

11. ISO TC 184/SC 4/WG 5, N 31 (P 4). Draft, 25 January 1992. STEP Development Methods: Resource Integration and Application Interpretation. W.F. Danner, Y. Yang.

12. ISO TC 184/SC 4/WG 1, N 329. General AEC Reference Model (GARM). April 1989. W.F. Gielingh.

13. IMPPACT ESPRIT Project 2165. Deliverable 108: EXPRESS description of the integrated information model, December 90.

14. ISO TC 184/SC 4/WG 1. Status: Draft. October 31, 1988. Annex D. Section 7: Ship Structural System Information Model.

15. NISTIR. A Proposed Framework for Product Data Modeling. William F. Danner. January 1990.

16. ISO TC 184/SC 4. DOCUMENT GPDM. DRAFT. Version 0.7., 10 January 1990. Generic Product Data Model (GPDM). W.F. Danner.

17.  ISO TC 184/SC 4. Document GPDR. DRAFT. Version 0.9a., 2 January 1991.
     Generic Product Data Resources for STEP. W.F. Danner.

18.  ISO TC 184/SC 4/WG 5, N 31 (P 4). Draft, 25 January 1992. STEP
     Development Methods: Resource Integration and Application Interpretation.
     W.F. Danner, Y. Yang.

19.  ISO TC 184/SC 4/WG 4, N 34 (P 5). Draft, 20 February 1992. Guidelines for
     the Development and Approval of STEP Application Protocols. Version 1.0.
     M. Palmer.

20.  ISO TC 184/SC 4/WG 1.DOCUMENT 3.2.2.5. DRAFT. Version 4.0.,
     18 May 1990. Application Reference Model for Ship Structural Systems.
     M. Gerardi.

21.  NIDDESC Ship Structure AP. Working Draft, Version 0.7, 12 April 1992.
     Submitted to ISO. M. Gerardi, P.L. Lazo, M.A. Polini.

22.  ISO TC 184/SC 4/WG 1. DOCUMENT 4.2.1 Working Draft, Version 1.1.,
     December 22, 1989. Reference Model for Distribution Systems.
     Douglas J. Martin.

23.  ISO DIS 10303-11: Industrial Automation Systems-Exchange of Product Model
     Data - Part 11: Description Methods: The Express Language Reference
     Manual, Doc. N14, ISO TC 184/SC4/WG5, April 1991.

24.  B. G. Wenzel, S. Müllenbach: EXPRESS Version 2, Requirements and Project
     Proposal, version 0.4, Working Draft, April 1992.

25.  P. Brun, A. Cazaux, Ph. Brèche, J. Lynch: Synthesis of the Specification
     Activity, Deliverable 3.2.3 of NEUTRABAS, ESPRIT Project 2010, WP3,
     October 1991.

26.  Objectives of the Part Library Working Group, working paper of ISO TC
     184/SC4/WG2, Doc. N43, February 1992.

27.  P. Harrow: Library Structures for STEP, working paper of ISO TC
     184/SC4/WG2, March 1992.

# Annex I:

## Steel Structure Representation Diagram

(adapted from CHART II.1, Del.3.1.1)

\* INFORMATION flow is coded by the last two columns on the right hand end of the lines involved only: first column identifies the "line number"; second column the DESTINATION line numbers.

### GLOBAL REPRESENTATION OF THE SHIP

#### BASED ON

| CONCEPTS | ATTRIBUTES | LINE | TO LINE |
|---|---|---|---|
| **LEVEL 0** | | | |
| Global entity: Ship | characteristics and main functions | | |
| **LEVEL 1** | **Whole ship representation (not included in NEUTRABAS)** | | |
| **LEVEL 2** | **Space organization** | | |
| **Level 2.1** | **Referential** | --- | **301** |
| Referential System: Transversal Longitudinal Horizontal | identification location | | |
| **Level 2.2** | **Surface organization** | --- | **302** |
| Major surfaces: Hull Decks Bulkheads | identification location geometrical surface definition boundaries | | |
| **Level 2.3** | **Volumic organization** | | |
| Volumes: Holds Engine compartment Accommodations Capacities | identification referential system geometric description (major surfaces nomenclature) characteristics of containment external solicitations | | |
| **LEVEL 3** | **Global perception of the steel structure** | | |
| Global entity: Steel structure | referential: gen. trihedral axes transv./long./horiz. partition geom. major surf. nomenclature struct. elements nomenclature weight, c. of g. | 301 302 | AAA 401 404 |

| CONCEPTS | ATTRIBUTES | LINE | TO LINE |
|---|---|---|---|
| **LEVEL 4** | **General partition of the steel structure** | | |
| **Level 4.1** | **Major surfaces partition** | **401** | |
| Major surface | identification | | |
| | joints nomenclature | 402 | 403,514 |
| **Level 4.1.1** | Joints definition | 403 | |
| | identification | | |
| | location/ contours definition | | |
| | extremes definition | | |
| **Level 4.2** | **Structural and spatial partition** | **404** | |
| Structural elements concepts | identification | | |
| | major surfaces nomenclature with: | | |
| | identification of associated | | 408 |
| | primary substructural elements | | |
| | corresponding prioritary | 405 | |
| | boundaries (ref. to surf. joints) | | 402 |
| | standards of relationship between | | |
| | second. and priorit. bounds. | | 402 |
| | location | AAA | BBB |
| | prefabrication blocks nomenclature | | 406 |
| | weight, c. of g. | | |
| **Level 4.2.1** | Spatial partition of the structural elements | 406 | 503,527 |
| Prefabrication blocks concepts | identification | | 504 |
| | location | BBB | |
| | boundary defin. (surface joints) | 407 | 402 |
| | weight, c. of g. | | |
| **Level 4.2.2** | Structural parts of the structural elements | 408 | |
| Primary substructure concepts | identification | | |
| | location | BBB | CCC,DDD |
| | plate sheets nomenclature | | 501 |
| | stiffening substr. nomenclature | | 526 |
| **LEVEL 5** | **General structure representation** | | |
| **Level 5.1** | **Shell plating** | **501** | |
| Plate sheets | identification | 502 | |
| | location | CCC | LLL,MMM |
| | inc. thickness orientation | 503 | |
| | prefab. plate sheets nomenclature | 504 | 507 |
| | boundaries nomenclature (surfaces) | | 506,520 |
| | plate assemblies nomenclature | | 505 |

| CONCEPTS | ATTRIBUTES | LINE | TO LINE |
|---|---|---|---|
| **Level 5.1.1** | Partition of the plate sheets | 505 | |
| Plates assemblies | identification | | |
| | boundaries nomenclature (surfaces | 506 | |
| | standard defin. (thickness, grade) | | 508 |
| | prefab. plates assemblies nomencl. | 507 | |
| | plates nomenclature | | 513 |
| *Level 5.1.1.1* | *Boundaries definition* | *508* | |
| Joints of plate assembies | identification | | |
| | standard definition: | | |
| |    welding strings  identif. | | |
| |    location | | |
| |    profiles | | |
| | strings nomenclature | | 509 |
| *Level 5.1.1.1.1* | *Partition of joints strings of plates assemblies* | *509* | |
| Joints strings (assemblies) | identification | | |
| | substrings nomenclature | | 511 |
| | substrings extremities nomencl. | | 510 |
| *Level 5.1.1.1.1.1* | *Definition of substrings extremities* | *510* | *512* |
| Substring extremity | identification | | |
| | standard | | |
| | location | | |
| *Level 5.1.1.1.1.2* | *Definition of the substrings* | *511* | |
| Substrings | | 512 | |
| | identification | | |
| | extremity identification | | |
| *Level 5.1.1.2* | *Partition of the plate assemblies* | *513* | |
| Plates | identification | | |
| | boundary nomenclature (surfaces) | 514 | 515 |
| |    and location | | |
| *Level 5.1.1.2.1* | *Boundaries definition* | *515* | |
| Joints of plates | identification | | |
| | standard definition: | | |
| |    welding strings  identif. | | |
| |    location | | |
| |    profile | | |
| | strings nomenclature | | 516 |
| *Level 5.1.1.2.1.1* | *Partition of joints strings of the plates* | *516* | |
| Joints strings (plates) | identification | | |
| | substrings nomenclature | | 518 |
| | substrings extremities nomencl. | | 517 |

| CONCEPTS | ATTRIBUTES | LINE | TO LINE |
|---|---|---|---|
| *Level 5.1.1.2.1.1.1* | *Definition of substrings extremities* | *517* | *519* |
| Substring extremity | identification<br>standard<br>location | | |
| *Level 5.1.1.2.1.1.2* | *Definition of the substring* | *518* | |
| Substring | identification<br>extremities identification | *519* | |
| **Level 5.1.2** | **Boundaries definition** | 520 | |
| Plate sheets joints | identification<br>standard definition:<br>   welding strings   identif.<br>     location<br>     profile<br>strings nomenclature | | 521 |
| *Level 5.1.2.1* | *Joints strings partition of plate sheets* | *521* | |
| Joints strings<br>(plate sheets) | identification<br>substring nomenclature<br>substring extremities nomenclature | | 523<br>522 |
| *Level 5.1.2.1.1* | *Definition of substrings extremities* | *522* | *524* |
| Substring extremity | identification<br>standard<br>location | 523 | |
| *Level 5.1.2.1.2* | *Definition of substrings* | | |
| Substring | identification<br>extremities identification | 524 | |
| **Level 5.2** | **Stiffening** | 525 | |
| Stiffening<br>substructure<br>element<br>  normalized<br>  composite | identification<br>location including<br>   orientation of std.<br>prefabr. stiff. nomenclature<br>stiffener nomenclature | DDD<br>526 | MMM,RRR<br>528<br>527,528 |
| **Level 5.2.1** | Partition of the stiffening<br>substructure | 527 | |
| Stiffener | identification<br>location<br>prefab. stiffener nomenclature<br>standard identification<br>connection butt joint nomencl.<br>string nomenclature (connection<br>  with shell plating) | RRR<br>528<br><br><br>529 | 529,530<br>535<br>534 |

| CONCEPTS | ATTRIBUTES | LINE | TO LINE |
|---|---|---|---|
| *Level 5.2.1.1* | *Standard content* | *530* | |
| Whole reference trihedron | definition | RRR | |
| Elementary pieces | identification reference trihedron contours definition thickness apertures (inner contours) | RRR | 546 |
| Elementary pieces arrangement | relative position nomenclature of strings between pieces | RRR | 531 |
| Definition of strings between pieces | identification standard library:      identifiaction            attributes strings extremities nomenclature substrings nomenclature | 531 | 534 532 533 |
| Definition of strings extremities | identification standards library:      identification            attributes location | 532 RRR | |
| Definition of substrings | identification extremities identification standards information | 533 534 | |
| *Level 5.2.1.2* | *Connection butt joint definition* | *535* | |
| Connection butt joint | identification standard identification location (where needed) | RRR | 536 |
| *Level 5.2.1.2.1* | *Standard content* | *536* | |
| Whole reference trihedron | definition | | |
| Elementary pieces (includes incidences of surrounding structure) | identification reference trihedron contours definition thickness apertures (inner contours) | RRR | 546 |
| Elementary pieces arrangement | relative position nomenclature of strings       between pieces | RRR | 537 |
| Definition of strings between pieces | identification standards library:      identification            attributes strings extremities nomenclature substrings nomenclature | 537 | 540 538 539 |

| CONCEPTS | ATTRIBUTES | LINE | TO LINE |
|---|---|---|---|
| Definition of strings extremities | identification<br>standard library:     identification<br>          attributes<br>location | 538<br><br><br>RRR | |
| Definition of substrings | identification<br>extremities identification<br>standards information | 539<br><br>540 | |
| *Level 5.2.1.3* | *Connection with shell plating* | | |
| String definition | identification<br>location<br>standards library:     identification<br>          attributes<br>substrings extremities nomencl.<br>substrings nomenclature | 541<br>RRR<br><br><br><br> | <br><br><br><br>542<br>543 |
| *Level 5.2.1.3.1* | *Definition of substrings extremities (connection with shell plating)* | *542* | |
| Substring extremities | identification<br>standards library:     identification<br>          attributes<br>location | <br><br><br>RRR | 544 |
| *Level 5.2.1.3.2* | *Definition of the substring (connection with shell plating)* | *543* | |
| Substring | identification<br>extremities identification<br>standards information | 544<br><br>545 | |
| **Level 5.3** | **Assemblies of apertures** | | |
| Aperture assembly | identification SE / PS / SS<br>identification<br>standard identification<br>location | 546<br><br><br>MMM | <br><br>547 |
| **Level 5.3.1** | Standard content | 547 | |
| Standard | contours definition | | |
| **LEVEL 6** | **Representation of particular details** | | |
| Foundations<br>Bilge wells<br>Lift trunks<br>Lowered spaces | will require specific concepts and attributes and are to be treated as special entities | | |

# Annex II:

## Applications Related to Ship Structures

### * Abbreviations

A: Applications
C: Categories
P: Predicates

See Section 3.4.1.4 for definitions.

### * Concept and General Requirements

- type of product (C)
- owner requirements (C)
- yard and fabrication requirements (C)
- quality level and administration (C)
- materials (C)
- standards (C)
- norms (C)
- regulations (C)
- codes (C)
- suppliers (C)
- ordering dimensions (C)
- limiting values and constraints (C)
- referentials (C).

Norms, referentials and other can be predicated at different stages along the life-cycle. Any element, or assembly of elementary components may be defined using a norm or conversely include a norm that defines other components or assemblies or attributes of those.

Likewise, any element may be used as a reference by others or may refer any other. Moreover, such reference may consist of a precise point, line or surface of that element.

### * Preliminary Design

Applications of this stage refer normally to the complete ship. However, for each ship type, some elements, generally taken as sub-structures or assemblies, are referred to for specific checks, calculations or other uses.

Therefore, the following list is to be construed as applicable to any type of element from the single item or part to the complete product or ship hull, although with different meanings varying with the complexity of the subject item.

- existence (P)
- speciality (C)
- alternatives (C)
- spatial (C):   outer dimensions (P)
                      location (P)
                      orientation (P)
                      topology (C)
                      geometry (P)
                      capacities (P)
- restrictions (C)
- obligations (C)
- physical (C): weights (P)
                      accessibility (C)
                      integration (C)   up- and down-stream
                      variability (C)
                      fixity (C)            support, connection
                      mobility (C)       operational
- functionality (C)
- calculations (C)
- representations (C)
- modifications (C)

## * Design and Engineering

- materials :   properties (P)
                     behaviour (C)
                     selection (C)
                     response (A)
- geometry :   referentials (C)
                     shape (P)
                     dimension (P)
                     contours (P)
                     lines (P)
                     area (P)
                     moments (P)
                     weight (P)
                     drawing, display (C)
- topology :    volumes (P)
                     lay-out (C)
                     interference (C)
                     interrelation (C)
                     dependances :   location (C)
                            connection (C)
                            sequencing (C)
                            usage (C)
                            functionality (C)
                            assembling (C)
                     access (C) :       openings
                            connections
                            voids

- computing :  capacity (C)
                   mechanical  response (C) :
                          strength (A)
                          vibration (A)
                          noise (A)
                          roughness (A)
                          protection (C)
                          shock (A)
                          impact (A)
                          fatigue (A)
                   operational response (C)
                   transferences (C) :          to others (A)
                   compound capabilities (C) :
                          structural (C)
                          other functions (C)
                   costs (C)
- references:   identification
                   drawings
                   lists
                   norms
- relations to equipment and outfitting (C).

## * Drawings and Specification

Calculations in general, being considered as a category of applications may be seen as the set of all computational. efforts leading to the development of the Preliminary, Contract and Detail Design Drawings as well as to the selection of all the data necessary to define the Ship Specification

A typical set of Drawings developed during the Contract Design Stage is listed in Ref 1 as follows :

- General arrangement : profiles and decks
- arrangement of particular spaces
- structural classification plans
- steel scantlings
- arrangement of machinery : plans and sections
- electrical system
- fire control by decks
- HVAC systems
- piping systems
- capacity planci
- curves of forms
- floodable lengths
- trim and stability booklets
- damage stability conditions

It is easy to show the importance of the hull structure in the development of all these documents.

On the other hand, the same Ref.1 lists the typical Sections that are included in a Commercial Ship Specification. That list must be considered as another Category of Applications for NEUTRABAS in what concerns the uses to be asked from the Ship Hull Structure

A non all-inclusive list of these Sections follows in order to show some areas where the hull structure is to be used or referred to:

- structural hull
- houses, bulkheads
- sideports, doors, hatches, manholes
- hull fittings
- deck coverings
- insulation, lining, battens
- kingposts, booms, masts, davits
- rigging and lines
- piping-hull system
- HVAC
- fire detection and extinguishing
- painting and cementing
- life saving equipment
- commissary spaces
- utility spaces and workshops
- furniture and furnishings
- plumbing and accessories
- protection covers
- miscellaneous equipment and storage
- name plates markings
- joiner work and interior decoration
- stabilization systems
- container stowage and handling
- main propulsion and auxiliary machinery
- main shafting, bearings and propeller
- F.O., water, steam air and exhaust systems
- ventilation, air conditioning and refrigeration equipt.
- cargo systems
- cargo hold systems
- tank systems
- compressed air system
- pumps, piping
- insulations and lagging
- ladders, gratings, floors, platforms, walkways, railinys
- workshops, stores and repair equipment
- hull machinery
- instruments, gage boards, mechanical
- spares, engineering
- electrical systems, generation, distribution, controls
- lighting
- interior communications

- deck, engine and stewards' equipment and tools
- centralized engine room and bridge
- planning and scheduling, plans, instruction books
- tests and trials
- interior security.

**\* Materials Acquisition**

- suppliers :     identification
                            qualifying
- raw materials :     plates
                            shapes
- consumables :     wire
                            electrodes
                            gas
- terms and conditions: pricing
                            delivery
- reception :     inspection
                            testing
- storing :     handling
- sub-contracting :  pre-fabrication

**\* Planning, Manufacturing, Assembling and Erection**

- planning :     sequencing
                            accessing
                            preparing :       planing
                                        scaling
                                        priming
- manufacturing :  handling :     moving
                                        rotating
                            operations :       cutting
                                        beveling
                                        burning
                                        forming
                                        plying
                                        attaching
                                        joining
                                        assembling
                                        pre-outfitting
                            ancillary :       scafolding
                                        lugging
                                        jigging
                                        strengthening
                                        patterns
                                        moulds
- costing
- production control
- testing :     watertightness
                            inspections

                                  quality assurance
                                  tolerances
                                  loading

- reporting
- painting
- fitting :       reinforcements
                                  attachments
                                  details
                                  supports
                                  equipment items
- assembling : positioning
                                  fixing
                                  joining
                                  checking
                                  testing
- erection :     moving
                                  hanging, rising
                                  reinforcing
                                  leveling
                                  fixing
                                  joining
                                  inspection
                                  scafolding
                                  accessing
                                  protecting
                                  painting
                                  attaching
- launching
- outfitting :   planning
                                  checking
                                  accessing
                                  preparation
                                  pipeing
                                  cableing
                                  hull fittings
                                  ductings
                                  accommodation
                                  foundations
                                  supports
                                  hangers
                                  penetrations
                                  attachments
                                  lining
                                  insulation
                                  protection
                        decoration

**\* Finishing, Delivery, Tests and Trials**

- finishing-up : checking
  completion
  repairing
  replacing
- tests
- trials
- inspections
- approvals
- certifying.

**\* Inspection, Maintenance and Repair; Modifications**

- identification
- isolation
- accumulation
- segregation
- re-definition
- inspections
- controls : dimensions
  elements
  quality
  functionality
- replacements
- conversions
- substitutions
- revisions
- calculations.

# Annex III:

# On the Use of Boundaries and References in the Design of Ships

## 1 Attributes of Referentials and Boundaries

### 1.1 Referentials

We use here the term "referential" to indicate something, or some entity which can be used for/as a "reference" by others. In this sense, a "reference" is the action and effect of referring to a referential , thus making adequate use of it.

- Hull Geometry References.- There is an open, though limited, set of referentials that are normally used to define the geometry of a ship hull by means of geometrical shapes and surfaces. Although it is customary to use one single origin and only one system of coordinate planes, such reference can vary and be so complex that it expands into multiple references as the design progresses, when advanced aCAD and CAL tools are applied.

  Depending on the way hull shapes are produced, different types of references are needed and used by the designer :

  > one and only one global, 3-D Cartesian coordinate system;
  > several reference systems, for several sections of hull;
  > points, lines and surfaces previously defined as support.

- Derivations from Moulded Hull Geometry.- The reason for one Moulded Surface or Reference is well grounded in traditional Hull Design and Lofting. As such, physical dimensions of hull materials are referred to, and many Naval Architecture calculations use such ideal geormetry as reference. The main applications using the Hull Moulded Surface or Definition are :

  - defining intersection lines given by Waterplanes, Transverse and Longitudinal Vertical planes
  - defining position lines for strakes and stiffeners situating points, openings and other accidents
  - subdividing the hull into spaces, holds, tanks, etc. by giving intersecting planar, cylindrical and polyhedrical surfaces

  References for points, lines and surfaces attached to the hull.

  During the process of defining subdivision surfaces, lines and points of reference on the Hull Moulded Surface or within it, references can vary as the designer pleases. Any defined point, line, plane or surfdce can be used as a further reference to another, next geometrical element. It is a free, non-limited, relative reference process.

References can be geometric, i e. to derive a geometrical relationship to a given referential. They can also be physical, i.e. to define physical elements referred to by the same referential. And this referential can be either geometrical or physical.

References to one geometric element may be used to identify, locate, derive or manipulate another element.

During the definition and calculation phases of the Hull Design Process, such references can be made at any time, and for ease of location, referentials are identified uniquely, either for permanent use or just for eventual references. This implies using living or current referentials as well as permanent" or "fixed" ones.

• Dimensional values are also used for reference. In this case, these values can take the role of parameters of the reference function that is established. Such are

>   lengths, breadths, depths, drafts;
>   camber, siding, deadrise, sheer;
>   thickness of plating, bend radii;
>   scantlings of rolled and built-up sections.

## 1.2 Boundaries

The term "boundary" is used in this article with the meaning of a geometric limit, surrounding or enclosure for a region of a geometrical space, be it a line, a surface or a volume. Therefore, the existence of such domain of a n-D space is bi-univocally related to the existence of its boundary that is to be (n-1)-D.

Types of boundaries defining sub-spaces are :

>   3-D spaces are enclosed by surface boundaries;
>   Surface boundaries are surrounded by linear boundaries;
>   line boundaries are limited by end-points.

Boundaries are used only to limit spaces: distances, surfaces, planar or general, and volumes. Therefore, boundaries are points, lines and surfaces and have one dimension less than the geometrical entity they limit or enclose.

Boundaries can be geometric, e.g. the moulded plane surface of a bulkhead. Boundaries can also be physical, e.g. the plating or stiffeners of the same bulkhead.

Boundaries can be used as referentials, both for geometric and physical references. Thus, the definition of the boundary has two points of view. In essence, a boundary is an "attribute" of a referential, as it is being used for reference to certain applications involving the concept of "limit", e.g. calculation of distances, areas or volumes.

Therefore, the same geometric entity may be used to support a physical entity, and both, geometrical and physical, may be used as referentials and as boundaries, for different purposes.

Boundaries exist only to be used as limiting edges of another geometric or physical entity. Therefore they mean a step forward in the applicability of just geometrical references.

Essential attributes of boundaries are :

geometry
location or topological relationship
physical dimensions, scantlings
material
fabrication features,
protection
access
restrictions of use
scope of application: support, protections, strength, etc
obligations and codes: rules, reinforcement, etc.

The role of physical entities as boundaries or referentials in the Design Process can be richer and more complex as the design progresses down to the Fabrication and Operation Phases. As an example, a bulkhead is used just as a geometric plane or distance to evaluate floodable lengths, whereas its plating, stiffeners and attached elements or openings may be used as references for other construction details related to such bulkhead or to the hold or the compartment where it belongs, and for which it is a boundary.

## 2  Treatment of Referentials and Boundaries in Ship Design

### 2.1  Concept Design Prefeasibility

Decisions on main dimensions and global values of ship features are adopted in this phase concerning :

mission : load and traffic
type and quality of vessel
speed and other service requirements
propulsion system
main dimensions
coefficients of form
main subdivisions
cargo handling
materials and equipment
Class Rules and other applicable codes
construction.

## 2.2 Preliminary Design Feasibility

• External Hull Geometry - Generally and traditionally simplified as a Moulded Surface, it is defined as a unique and continuous combination of surfaces, enclosing the hull volume, that intersect or join one another at common edges and joints.

Hull lines and body plan, can be produced :

- from scratch, creating profile, significant sections and cross sectional area curve to define all lines;
- using previous hull lines and transforming them;
- defining main-wire model lines, planar and spatial;
- sculpturing the hull moulded surfaces from a first trial;
- defining orthogonal plane lines and spatial references
- combining data-point coordinates and faired lines
- adapting 3-D surfaces or patches onto pre-defined lines
- interpolating values into pre-defined lines, forms and other mathematical definitions of shapes.

Hull surfaces can be defined :

directly, mathematically or sculpturally;
using pre-defined 2-D and 3-D lines;
using lines interpolated through data-points;
parametrically, from existing models and data-banks

Applications of external hull geometry include :

hydrostatic characteristics of floating body;
estimate of speed and powering;
calculation of flodable length;
use as a boundary for internal subdivision;
definition of outer shell components;
location of appendages and hull openings;
positioning of main equipment and machinery onboard

• Internal Hull Geometry - Along with the Hull Design Definition Process, spaces within the hull are segregated by internal hull subdivision, achieved by definition of planes and surfaces, and by identification of surfaces and their edges.

Definition of subdivision boundaries can be achieved by :

using points and lines defined in the outer hull surface
locating them by using geometrical referentials.

Interior boundaries sub-divide spaces in order to:

define attributes and functions of sub-spaces;
calculate lengths, areas. volumes, centres, moments

      evaluate loadings onto the physical boundary;
      define connectivity : passages, openings, insulation.

Subdivision into sub-spaces permits :

      locate elements and equipment in each sub-space;
      distribute cables, pipe- and duct-work;
      check for interferences and accessibility;
      calculate weights and centres, for loading conditions
      produce cargo tables, sounds and ullages;
      simulate or model damage stability;
      locate access openings and comrnunication means.

• Physical Interior Boundaries - Exist as entities that can be derived from previous geometrical boundaries upon which some physical attributes have been added, such as :

      scantlings and material.

Applications along the Ship Design Process refer to Physical Boundaries at, and beyond the Contract Design Stage.


## 2.3  Contract Design/Specification Phase

This phase of the Ship Design Process is mostly concerned with the calculation and selection procedures that assess the tentative values of features and dimensions established in the Preliminary Design using simple evaluation approaches. This phase provides the support for defining individual elements and pieces of equiprnent and systems to be ordered or fabricated. The results of this Design Phase are recorded in the Contract Specification.

Referentials and Boundaries used at the Contract Design Phase are predominantly the same as required to support the development of the Preliminary Design Phase. However, some of the elements that are defined in the Preliminary Design Phase are to be used for extensive calculations during the Contract Design Phase.

Regarding the Hull Structure, Contract Design deals with the modelling and scantlings of the structure. It begins with the configuration or arrangement of structural members, and solves for the physiscal dimensions of stiffeners and connections, plus other structural elements necessary for the construction of the hull.

References are used as global, local and detail :

• global, to deal with :
      tanks
      holds
      decks and platforms
      superstructures
      deckhouses

- local, to calculate or design
  - side shell
  - transverse and longitudinal bulkheads
  - gross panels of decks and platforms
  - solutions of superstructures and deckhouses
  - pillars and masts
  - foundations

- detailed, to assess and decide on
  - connectivity of elements
  - supporting conditions
  - structural discontinuities
  - crossing, joining and stiffening of elements
  - use of yard norms

## 2.4 Engineering/Detail Design

In this phase, all details necessary for the construction of the ship are defined and calculated. This includes any definition of geometry and other manufacturing pararneters, that are to be used by a fabrication process, generally of the CAM type, or combined with CAD systems.

The outcome of this phase are detailed part drawings to be used in the steel fabrication shop, either manually or at N/C stations by a CAM system.

For the definition of structural parts and elements, several references are normally required at the same time. One single element, e g. a bracket or a stiffener, can be defined "directly" using a geometrical CAD language, or else it can be "indirectly" defined using a "macro procedure" or "norm" for which its parameters are given. In both cases, dimensional and positional values are used that require making references to other points, lines, surfaces or elements in the same region.

Furthermore, it is generally assumed during the CAD detailing process that dimensional and positional references are stated by the use of parametric constants, whose values are solved by some calculation procedure or algorithm, in such a way that references can be nested to one another, and be relative to some other, and a complex network of references may be existing while the numerical solution of all values is achieved.

Therefore, an attribute to be included in Referentials is the reference to other referentials, and the type of reference that is involved in that reference.

References can be made simultaneously to global referentials and to local or detail ones. Thus, in order to integrate a plate chock or clip into a fabrication assembly, it should be related to its "mother part", and this one to the assembly, but for integration of centres of gravity, and for maintenance of the planning and control programme, the actual global position that refers to the ship as a whole has to be supported at the same time.

Therefore, any single part or component of the Product should have to be allowed to maintain a free, although reasonably limited number of references to different referentials. Despite that the values of some of the parameters of certain references are solved at different phases of the Design-Fabrication process, and are results for other uses, and not applicable for the design proper.

## 3  A Sample Application to Ship Design

### 3.1  Global Coordinate System

The ship hull, as required by hydrodynamic and other mission related criteria is defined by a set of geometrical elements which are referred to a Global Referential System, generally defined by three Cartesian coordinate planes :

plane OXY :     horizontal, contains the Base Line of the hull
plane OXZ :     vertical, longitudinal plane of symmetry, center plane
plane OYZ :     vertical, transverse, located at one of :
                  -   aft perpendicular
                  -   foreward perpendicular
                  -   midship section

First Level of Fabrication generally uses this Referential for reference.

### 3.2  First Level Coordinate System

A First Level Coordinate System is used by all families of Ship hull components of the First Level. Each family uses its own Referential.

Families of the First Level are

-   elements that are independent, e.g. assemblies
-   surfaces that are, or can be consiodered as independent, e.g a tank on a deck
-   independent equipment, e.g. fire-fighting guns, on decks
-   other elements that are referred to Global Referential.

The components of the Second Level parts or elements are referred to these First Level Coordinate Systems .

Every First Level Referential System is defined by its origin and orientation of the three axes referred to the Global Reference System.

## 3.3 Nth-Level Coordinate System

A Nth-Level Coordinate System is that system defined in association with a Nth-level family of hull components, which is used as a referential by the (N+1)th-level families of components. Every Nth-Level Coordinate System is defined with reference to one (N-1)th level Referential.

## 3.4 Attributes of a Referential

- Identification : it should include the Level of Referential
- Referential : of upper level to which it is attached or related
- Identification : it should include the Level of Referential
- Origin : coordinates in the Referential to locate it
- Axes : orientation with respect to same Referential
- Comments : to explain the reasons to define it

## 3.5 Examples

### 3.5.1 Positioning the Flanges of Piping-Work in a Bulkhead

Generally, flanges are specified to be located at x cm off centers, along a certain line.

This line is positioned within the bulkhead. If the bulkhead is sub-divided into more than one fabrication ascembly, more than one different local referntial may be required in order to position all flanges.

Assemblies are further referred to bigger fabrication units. These, in turn are located with reference to the Hull Global System, or relative to another assembly, or to a distinct element of it.

### 3.5.2 Locating a Connecting Flange in a Pump

Let us position ourselves at the lowest level of reference and follow all the way upstream up to the highest referential level, the only Global one.

As an example, we take the bolt holes at the flange of foaming liquid pump in a rescue and fire-fighting vessel and let it be at the 4th level of reference.

a. To bore the holes these have to be positioned with a reference to an intrinsic referential in the flange, where a system (04-1;X4-1,Y4-1,Z4-1) of coordinate axes is defined. The position of each hole center is referred with respect to this system, in Cartesian or polar coordinates.

b. The flange has to be attached to the pump, and therefore it has to be located with reference to this pump using some system of axes as a referential.

This referential (03-1;X3-1,Y3-1,Z3-1) is quite different from another referential, namely the raw plate edges, that is needed for cutting the flange out from a nested plate, which will be at the same level 3 for the flange, but could be just level 1 for the production phase.

c. The pump has to be bolted on its foundations, in the machinery room. Installation of the pump is normally done by positioning it with reference to a certain mark made on the foundation, e.g. at its centerline. Here, a new referential system is located (02-1,X2-1,Y2-1,Z2-1) with respect to which the referential system attached to the pump (03-1) is positioned and oriented.

d. On the other hand, when the pump foundation is welded onto the double bottom of the machinery space, some reference has to be taken to the hull structure, e g.frame 18, or one bulkhead. Therefore. the referential system (02-1) would have to be referred to such new referential, (01-2;X1-2,Y1-2,Z1-2).

   If the the pump was incorporated at the pre-outfitting phase of a steel assembly, this referential (01-2) could have been the same used to position the assembly at the erection herth relative to other assemblies or to a referential related to the boundaries of the assembly (01-1:X1-1.Y1-1.Z1-1)

e. Finally, the whole assembly is positioned with reference to a global referential system, generally the only one fixed system of reference for the whole ship. (OG;XG,YG,ZG).

## Annex IV:

## List of NEUTRABAS Structural System Entities

(not including entities owned by other models)

activities dependency
activity
activity definition
activity function
assembly part
commercial stiffeners
composite stiffeners
concept
connecting part
connections and joints
connection resource activity
control definition
cost control activity
cost control definition
definition concept
design access lightening hole
design access opening
design activity
design calculation activity
design connection type
design cutout hole
design definition
design drawing activity
design hatch opening
design hull
design large opening
design lift opening
design opening type
design plate
design plate assembly
design plate sheet
design primary substructure
design product
design shell door
design stair opening
design standard opening
design standard connection type
design stiffener
design stiffening element
design stiffening element assembly

design structural element
design structural opening
design system penetration hole
elementary function
elementary part
feature part
function definition
function model
general function
global activity
joint
local activity
management definition
material
miscellaneous
non-parametrized standard library
non-standard
outer contour
parametrized standard library
planned definition
planning activity
planning definition
planning model
plate_sheet joint
plate joint
presentation concept
product
product definition
product function
production activity
production assembly
production assembly activity
production compartment
production connecting part
production connections
production definition
production engineering activity
production engineering assembly
production engineering assembly activity
production engineering compartment

production engineering connecting part
production engineering connections
production engineering cutting activity
production engineering definition
production engineering flange
production engineering hull
production engineering joint
production engineering opening
production engineering painting activity
production engineering plate
production engineering prefabricated block
production engineering prefabricated sub-block
production engineering product
production engineering stiffener
production engineering sub_assembly
production engineering web
production flange
production function
production hull
production opening
production painting activity
production plate
production prefabricated block
production prefabricated sub-block
production product
production stiffener
production sub_assembly
production web
production welding activity
quality control definition
quality control model
reference part
representation concept
resource
space part
standard
standards library
system
system definition
transition part

## Annex V:

## NIAM Symbolic Notation and Sample Applications

The NEUTRABAS project like many other early STEP activities, which began before mature versions of EXPRESS became available, relied on the Nijssen Information Analysis Method (NIAM) [4] as the main methodology for developing and documenting its initial product models. The later conversion of NIAM models into more strictly formalized EXPRESS models was a necessary follow-up step, but did not present any grave difficulties.

NIAM uses a graphical symbolic notation to represent its information models. The notation refers to the following basic NIAM modeling constructs:

Objects
Roles between objects
Object and role constraints

The following is a list of relevant symbol definitions used in the context of the NEUTRABAS product model. This list is based on an informal communication received in 1989 through the courtesy of James Turner, The University of Michigan, and reproduced here with his permission.

### Object

Objects are tangible or abstract entities in an enterprise.

NOLOT - Non-Lexical Objectives represent a set of conceptual entities (without value assignment) having common properties. The symbol for a NOLOT is a solid circle containing the NOLOT name.

LOT - Lexical Objects represent a set of values of an entity, such as names and properties. The symbol for a LOT is a dashed circle containing the LOT name.

MODEL - The main NOLOT of the model. The remainder of the model usually supports its definition. The symbol for a MODEL is a heavy dashed circle containing the MODEL name.

This is a NIAM extension.

SUBMODEL - A token, or a place holder for a model. The expanded model is in another location in the same document. The symbol for a SUBMODEL is a circle containing the SUBMODEL name circumscribed in a square. This is a NIAM extension.

SUBMODEL - A token, or a place holder for a model. The expanded model is in another document. The symbol for a SUBMODEL is a circle containing the SUBMODEL name circumscribed in a black square. This is a NIAM extension

## Role

The relationship or association between two objects is called a ROLE. Role names are read as A-R1-B and B-R2-A, or members of A "play" role R1 with members of B and members of B play role R2 with members of A. Roles act as a relation between the members of A and B:

$$R1: \ A \rightarrow B$$
$$R2: \ B \rightarrow A$$

The set of occurrences of role R1 ($O_{R1}$) is equal to the subset of the cartesian product of A and B for which the role A-R1-B is true.

A role is shown as a divided box attached to the affected objects with solid lines. The role names, or phrases, are written either inside each box or outside the box and attached with a leader line.

If one of the role names is omitted, the missing co-role is assumed to be the inverse of the existing role.

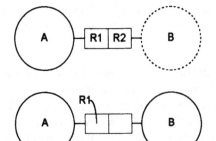

BRIDGE - A role between a NOLOT and a LOT

IDEA - A role between two NOLOTs.

## Object Subtypes

Object subtyping is a method to describe the characteristics of the subsets of a NOLOT

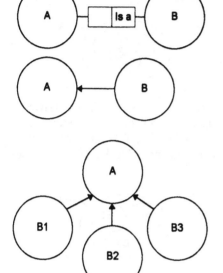

The "IS A" role is a common relationship between NOLOTs. As a result, a special symbol - a directed line segment - is provided. The arrow from B to A designates object B as a SUBTYPE of supertype object A, or set B is a SUBSET of set A.

SUBTYPE, SUBSET - B1, B2 and B3 are SUBSETS of SUBTYPES of A. Each member of A may be a member of B1, B2, B3, or any other subset of A. Or a member of A may be a member of any combination of B1, B2 and B3.

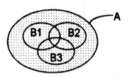

## Subtype Constraints

Subtype constraints are rules which restrict the division of a NOLOT into subsets. Subtypes are shown as a line connecting all affected subtype lines (arrowhead) with a circled letter superimposed. The letter designates the type of constraint.

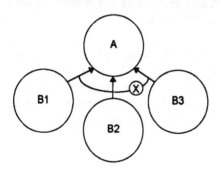

MUTUAL EXCLUSION - Each member of A can be a member of B1, B2, B3, or another subtype of A. B1, B2, and B3 are disjoint.

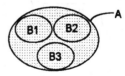

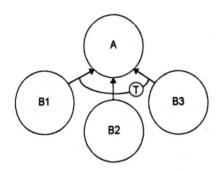

TOTAL - Each member of A must be a member of B1, B2, or B3; there are no other subtypes of A. Each member of A can be a member of more than one of the subtypes. B1, B2, and B3 may intersect.

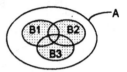

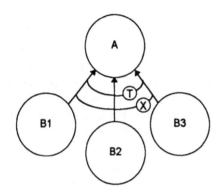

TOTAL MUTUAL EXCLUSION - Each member of A can either be a member of B1, B2, B3; There are no other subtypes of A. B1, B2, and B3 are disjoint.

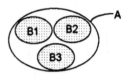

## Cardinality Constraints

Cardinality constraints designate the quantities of objects and roles allowed in a role.

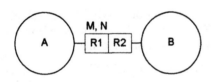

ROLE CARDINALITY - is shown as a minimum and maximum number above the affected role. Here, members of set A play between M and N roles R1 with members of B.

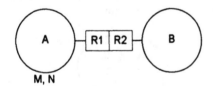

OBJECT CARDINALITY - is shown as a minimum and maximum number placed outside the affected object. Here, between M and N members of set A play role R1 with members of B.

## Idea Constraints

Idea constraints are restricting rules on roles between NOLOTs and are used to define the semantics of the relationships between objects. Idea constraints are divided into UNIQUENESS and TOTAL constraints.

A UNIQUENESS constraint is drawn as a line above or below the role. The line may or may not have arrowheads drawn at both ends.

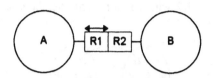

UNIQUENESS - Each member of A plays role R1 with zero or one member of B. Each member of B plays Role R2 with zero, one, or many members of A.

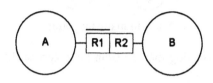

This constraint defines R1 as an identifying role of A.

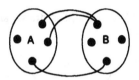

Many to one mapping

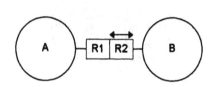

UNIQUENESS - Each member of B plays role R2 with zero or one member of A. Each member of A plays Role R1 with zero, one, or many members of B.

This constraint defines R2 as an identifying role of B.

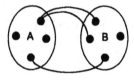

One to many mapping

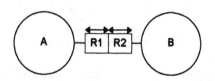

UNIQUENESS - Each member of A plays role R1 with zero or one member of B. Each member of B plays role R2 with zero or one member of A.

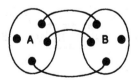

One to one mapping

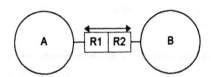

UNIQUENESS - Each member of A plays role R1 with zero or many members of B. Each member of B plays role R2 with zero or many members of A.

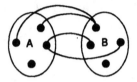

Many to many mapping

A TOTAL constraint is drawn as a "V" intersecting the line from the object to the role box.

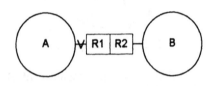

TOTAL - Each member of A plays role R1 with one or many members of B.

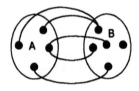

**Combined Uniqueness and Total Constraints**

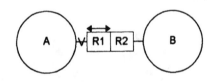

Each member of A plays role R1 with one and only one member of B. Each member of B plays Role R2 with zero, one, or many members of A.

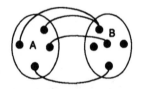

Total many to one mapping

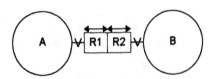

Each member of A plays role R1 with one and only one member of B. Each member of B plays role R2 with one and only one member of A.

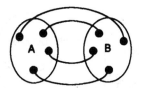

Total one to one mapping

The use of NIAM in NEUTRABAS is illustrated by the following NIAM diagrams. The examples refer to the following NEUTRABAS modeling objects, whose natural language descriptions are also included except for the first one.

1. Ship Design Information Model
2. Ship Design Definition
3. Ship Design Product
4. Ship Design Structural Element
5. Ship Production Engineering Product
6. Ship Production Engineering Prefabricated Block

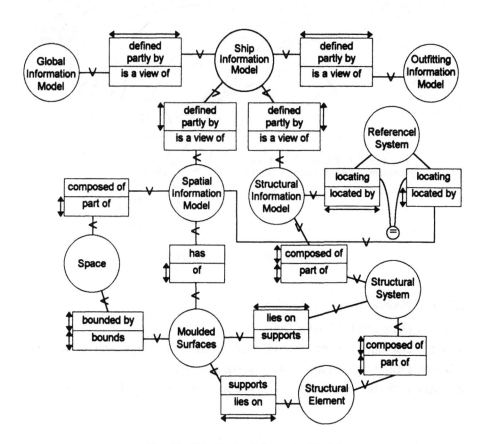

**Fig. 66** Ship Design Information Model

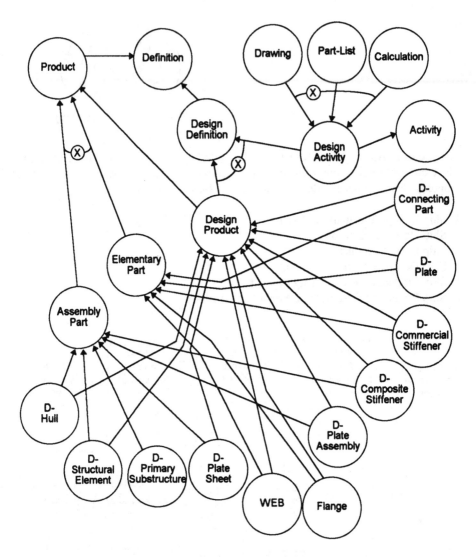

**Fig. 67**  Ship Design Definition

**ENTITY NAME: Design_Definition,**

a Design_Definition is a kind of Definition,

a Design_Definition may be either a Design_Product or a Design_Activity or other,

a Design_Definition must have exactly one design_type,

a Design_Definition must be identified by exactly one design_code.

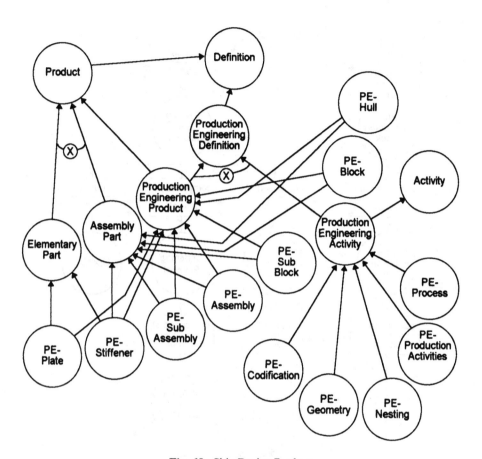

**Fig. 68**   Ship Design Product

**ENTITY NAME: Design_Product,**

a Design_Product is a kind of Design_Definition,

a Design_Product is a kind of Product,

a Design_Product may be either a
  D_Plate or a D_Commercial _Stiffener or a D_Connecting_Part or a D_Flange or a
  D_Web or a D_Hull or a D_Structural_Element or a D_Primary_Substructure or a
  D_Plate_Sheet or a D_Plate_ Assembly or a D_Composite_Stiffener or a
  D_Opening or a D_Connection or a D_Joint or a D_Space or a D_Compartment or
  other,

a Design_Product must have exactly one design_product_type,

a Design_Product must be the technical solution of exactly one Product_Function,

a Design_Product must have exactly one design cost.

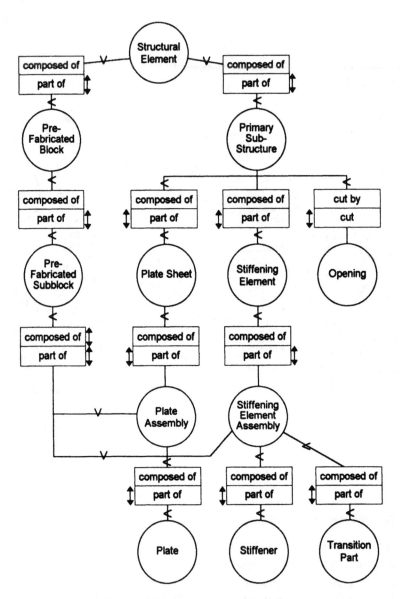

**Fig. 69** Ship Design Structural Element

**ENTITY NAME: D_Structural_Element,**

a D_Structural_Element is a kind of Design_Product,

a D_Structural_Element is a kind of Qssembly_Part,

a D_Structural_Element must be the component of exactly one parent D_Hull,

a D_Structural_Element must have exactly one structural_element_type
(double_bottom, bulkhead, deck),

a D_Structural_Element must be composed of at least two other
D_Primary_Substructure,

a D_Structural_Element must lie on one or more Moulded_Surface.

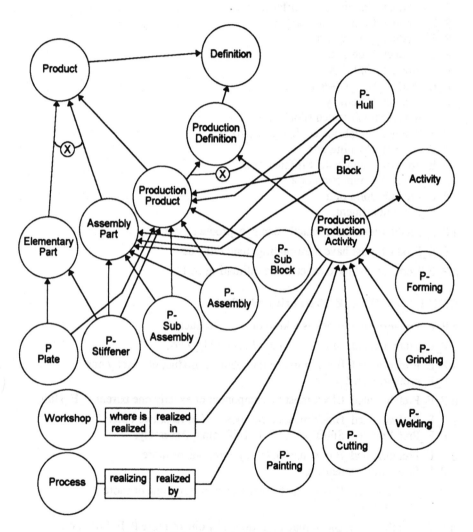

**Fig. 70** Ship Production Engineering Product

**ENTITY NAME: Production_Engineering_Product (P_E_Product),**

a P_E_Product is a kind of Product_Engineering_Definition,

a P_E_Product is a kind of Product,

a P_E_Product may be either a
  P_E_Product_Plate or
  P_E_Product_Commercial_Stiffener or a
  P_E_Product_Connecting_Part or a
  P_E_Product_Flange or a
  P_E_Product_Web or a
  P_E_Product_Hull or a
  P_E_Product_Sub_Assembly or a
  P_E_Product_Qssembly or a
  P E Product Prefabricated Block or a
  P_E_Product_Prefabricated_Sub_Block or a
  P_E_Product_Opening or a
  P_E_Product_Composite_Stiffener or a
  P_E_Product_Connection or a
  P_E_Product_Joint or a
  P_E_P_Compartment,

a P_E_Product must have exactly one p_e_product_type,

a P_E_Product must have a estimating_material_cost,

a P_E Product must have exactly one estimating cost.

**ENTITY NAME: P_E_Prefabricated_Block,**

a P_E_Prefabricated_Block is a kind of P_E_Product,

a P_E_Prefabricated_Block is a kind of Assembly_Part,

a P_E_Prefabricated_Block must be the planned product of exactly one
  P_E_prefabriocated_Block,

a P_E_Prefabricated_Block must be component of exactly one parent P_E_Hull,

a P_E_Prefabricated_Block must be composed of at least two other
  P_E_Prefabricated_SubBlock or P_E_Stiffening_Assembly,

a P_E_Prefabricated_Block may be cut by zero, one or more
  P_E_Structural_Opnening,

a P_E_Prefabricated_Block may be penetrated by zero, one or more
  P_E_Cutout_Hole,

a P_E_Prefabricated_Block may be stiffened by one or more P_E_Stiffener,

a P_E_Prefabricated_Block must be connected to other P_E_Product by one or more
  P_E_Connecting_Part,

a P_E_Prefabricated_Block must be joined to other P_E_Product by one or more
  P_E Joint

a P_E_Prefabricated_Block has exactly one thickness,

a P_E_Prefabricated__Block must have exactly one thickness_orientation.

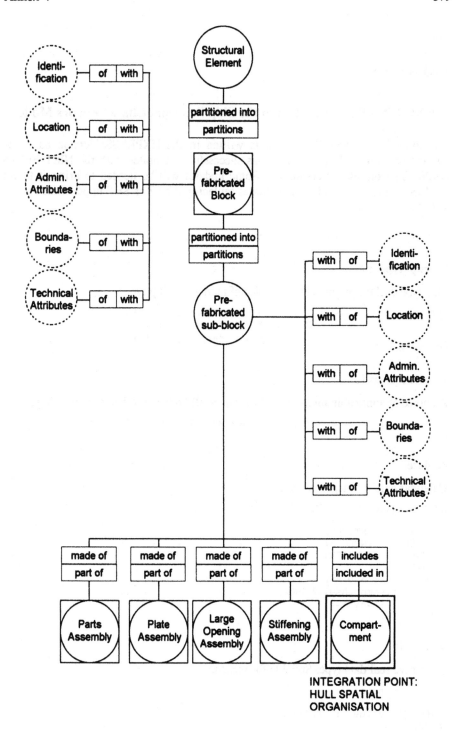

**Fig. 71**  Ship Production Engineering Prefabricated Block

# Annex VI:

# EXPRESS Codes

## 1 EXPRESS Representation of the Ship Principal Characteristics Model

The following EXPRESS schema is written in the EXPRESS Version based on STEP Doc. N 496 (July 1990). It was successfully compiled with the NIST FED-X EXPRESS Compiler, Version 1.1. Some updates will be needed to bring it to the final IS EXPRESS Level. This includes replacing # by ? as SET upper limit as well as OR by ANDOR.

```
SCHEMA    ship_principal_characteristics_model;
```

### 1. ship_type_name
This type declaration provides a choice of names for ship types.

```
TYPE ship_type_name = ENUMERATION of (Container_ship,
                      RORO_ship, Short_sea_ferry,
                      Icebreaker, Trawler, Tanker);
END_TYPE;
```

### 2. name_of_rule
A choice of applicable international (or national) bodies of rules for ship safety.

```
TYPE name_of_rule = ENUMERATION of (MARPOL, SOLAS, IMO);
END_TYPE;
```

### 3. date
Calendar date in integer syntax.

```
ENTITY   date;

   year   :   INTEGER;
   month  :   INTEGER;
   day    :   INTEGER;
   WHERE
   w1 : year > 0 AND
        {1<=month<=12} AND
        {1<=day<=31};
END_ENTITY;
```

### 4. metric
A positive real data type. Suitable for measures.

```
ENTITY   metric;

  default_value  :  REAL;
WHERE
  w2 : default_value >= 0;
END_ENTITY;
```

```
(* This ENTITY is meant to provide an ENTITY-data-
   type which can be used for all attributes which are
   of type REAL and positive, it may be dropped later
   since TYPEs can now be provided with local
   constraints.        *)
```

```
(* GENERAL PURPOSE AND AUXILIARY ENTITIES*)
```

## 5. ship-type

### An auxiliary entity designating functional ship type.

```
ENTITY ship_type;

   functional_type : ship_type_name;
END_ENTITY;
```

## 6. ship_rules

### An auxiliary entity for the name of the applicable ship safety rules.

```
ENTITY ship_rules;

   rule_name : name_of_rule;
END_ENTITY;
```

## 7. yard

### Name of the building shipyard.

```
ENTITY yard;

   yard_name : STRING;
END_ENTITY;
```

## 8. classification_society

### Name of the classification society to whose rules the ship is built and classified.

```
ENTITY classification_society;

   Class_name : STRING;
END_ENTITY;
```

## 9. person

### A person's name.

```
ENTITY person;

   person_name : STRING;
END_ENTITY;
```

## 10. line

### The length of a reference line.

```
ENTITY line;

  line_length : REAL;
END_ENTITY;
```

## 11. area

The area of a reference plane.

```
ENTITY plane;

  plane_area : REAL;
END_ENTITY;
```

## 12. weight

The weight of a weight category or the magnitude of a local weight (load).

```
ENTITY weight;

  total        : REAL;
  local_weight : REAL;
END_ENTITY;
```

## 13. main_engine

Engine type by manufacturer's name or code.

```
ENTITY main_engine;

  typpe : STRING;          --the use of ENUMERATION types is
END_ENTITY;                --probably more adequate here.
```

## 14. gear

The type of the gear by name or code.

```
ENTITY gear;

  typpe : STRING;          --same
END_ENTITY;
```

## 15. auxiliary

The type of auxiliary by name or code.

```
ENTITY auxiliary;

  typpe : STRING;          --same
END_ENTITY;
```

## 16. gen_set

The type of generator set by name or code.

```
ENTITY gen_sets;

  typpe : STRING;          --same
END_ENTITY;
```

## 17. rudder

The type of rudder by name or code.

```
ENTITY rudder;

   typpe : STRING;          --same
END_ENTITY;
```

## 18. propulsion_system

The type of propulsion system by name or code.

```
ENTITY propulsion_system;

   typpe : STRING;          --same
END_ENTITY;
```

## 19. screw

The type of propeller by name or code.

```
ENTITY screw;

   typpe : STRING;          --same
END_ENTITY;
```

## 20. prop_aids

The type of auxiliary propulsion device by name or code.

```
ENTITY prop_aids;

   typpe : STRING;          --same
END_ENTITY;
```

## 21. thruster

The type of thruster by name or code.

```
ENTITY thruster;

   typpe : STRING;
END_ENTITY;
```

## 22. deck_machinery

The type of deck machinery by name or code.

```
ENTITY deck_machinery;

   typpe : STRING;          --same
END_ENTITY;
```

## 23. crane

The type of crane (on deck) by name or code.

```
ENTITY crane;

   typpe : STRING;           --same
END_ENTITY;
```

## 24. hatch

The type of hatch system by name or code.

```
ENTITY hatch;

   typpe : STRING;           --same
END_ENTITY;
```

## 25. windlass

The type of windlass by name or code.

```
ENTITY windlass;

   typpe : STRING;           --same
END_ENTITY;
```

## 26. rescue_device

The type of rescue device by name or code.

```
ENTITY rescue_device;

   typpe : STRING;           --same
END_ENTITY;
```

## 27. nautical_device

The type of nautical device by name or code.

```
ENTITY nautical_device;

   typpe : STRING;           --same
END_ENTITY;
```

## 28. computer

The type of shipboard computer by name or code.

```
ENTITY computer;

   typpe : STRING;           --same
END_ENTITY;
```

## 29. compartment

A shipboard space characterized by location, volume capacities and name.

```
ENTITY compartment;

   location     : STRING;
                           --or reference to coordinate system
   volume_total : REAL;
   volume_gross : REAL;
```

```
   volume_grain : REAL;
   volume_bale  : REAL;
   identifier   : STRING;
END_ENTITY;
```

## 30. door

**The type of door system by name or code.**

```
ENTITY door;

   typpe : STRING;
END_ENTITY;
```

## 31. equation

**An expression written as text string.**

```
ENTITY equation;

   expression : STRING;
END_ENTITY;
(* MAIN ENTITIES OF THE SCHEMA *)
```

## 32. design_requirements

**A set of owner's and operational requirements, regulatory and transport capacity requirements.**

```
ENTITY  design_requirements;

   ship_functional_type        :  STRING;
                            --e.g. drill ship, short-sea-ferry
   operation_area              :  STRING;
                                         --e.g. worldwide
   restrictions_acc_to_ports : SET  [0:#] of  STRING;
                                     --breadth_max, length_oa
   restrictions_acc_to_locks : SET  [0:#] of  STRING;
   restrictions_acc_to_bridges : metric;
                                   --max. allowable height
   transportation_facilities_required : SET [0:#] of STRING;
   transportation_capacities_required : SET [0:#] of STRING;
   self_unloading_facilities :  LOGICAL;
   classification_required   :  STRING;
   acc_to_rules              :  SET [0:#] of  STRING;
                                      --MARPOL, SOLAS etc.
   speed_required            : metric;
   range_required            : metric;
   manning_required          :  INTEGER;
   no_of_passengers_assigned :  OPTIONAL INTEGER;
UNIQUE
   u1:classification_required;
END_ENTITY;
```

## 33. general_characteristics

**A supertype of many classes of general characteristics of the ship.**

```
ENTITY  general_characteristics

  SUPERTYPE of ( ONEOF (general_description,
```

```
                             main_dimensions,
                             weight_summary,
                             machinery,
                             propulsion,
                             freeboard,
                             cargo_and_tank_capacities,
                             crew_and_passenger_capacities,
                             cargo_access_and_auxiliary_eqmt,
                             factors,
                             delivery_information));
END_ENTITY;
```

## 34. general_description

A set of general information identifying the shipbuilding project.

```
ENTITY  general_description

  SUBTYPE of (general_characteristics);

  project_yard      :   STRING;
  project_id        :   STRING;
  classified_acc_to :   SET  [1:#] of  STRING;
                                      --classification society

  classification_id :   STRING;
  no_of_rules       :   INTEGER;
 .ordered_by        :   STRING;
  order_date        : date;
  tender_date       : date;
  person_responsible :  STRING;
  project_status    :   STRING;
                                   --comments, approvals etc.
  general_arrangement_dwg_number :  STRING;
  framing_type      :   STRING;
                                 --longitudinal or transversal
  ship_status       :   OPTIONAL STRING;
                                    --e.g. in operation
UNIQUE
  u2:project_id, classification_id, order_date, tender_date,
      general_arrangement_dwg_number, framing_type;
WHERE
  w3:order_date <= tender_date;
  w4:no_of_rules >= 0;
END_ENTITY   ;
```

## 35. main_dimensions

A collection of ship main dimensions (measures, distances, scantlings) and other overall characteristics of the ship.

```
ENTITY  main_dimensions

  (* The terminology used is in accordance with DIN ISO 7462  *)

  SUBTYPE of (general_characteristics)
  SUPERTYPE  of (main_dimension_ratios);

  length_oa : metric;
  length_oa_submerged : metric;
  length_pp  : metric;
  length_wl : metric;
  breadth_moulded : metric;
```

```
      breadth_max : metric;
      no_of_decks : INTEGER;
      no_of_non_continuous_decks :   INTEGER;
      depth_to_main_deck : metric;
      depth_to_2nd_deck : OPTIONAL   metric;
      draught_moulded : metric;
      draught_aft : metric;
      draught_fwd : metric;
      draught_amidships : metric;
      draught_max : metric;                        --summer/tropical
      displacement_draught_moulded : metric;
      displacement_draught_max : metric;
      deadweight_draught_moulded : metric;
      deadweight_draught_max : metric;
      shell_plating_factor : metric;
      height_of_rounding : metric;
      keel_thickness : metric;
      bilge_radius : metric;
      speed_operating :   REAL;
      speed_trial_actual :   REAL;
      radius_of_action :   REAL;
      water_plane_area : metric;
      midship_section_area : metric;
      bulbous_bow :   LOGICAL;
      tonnage_system :   STRING;
      gross_tonnage :   INTEGER;
      net_tonnage :   INTEGER;
DERIVE
      trim :   REAL  := draught_fwd-draught_aft;
WHERE
      w5:length_oa >= length_pp;
      w6:length_oa_submerged >= length_wl;
      w7:breadth_moulded < breadth_max;
      w8:depth_to_main_deck >= depth_to_2nd_deck;
      w9:draught_max >= draught_moulded;
      w10:speed_trial_actual >= speed_operating;
      w11:speed_trial_actual > 0;
      w12:speed_operating > 0;
      w13:radius_of_action > 0;
      w14:displacement_draught_moulded <=
          displacement_draught_max;
      w15:deadweight_draught_moulded <=
          deadweight_draught_max;
      w16:no_of_decks >= 1;
      w17:no_of_non_continuous_decks >= 0;
      w18:gross_tonnage > 0;
      w19:net_tonnage > 0;
END_ENTITY;
```

## 36. main_dimension_ratios

Dimensionless ratios of principal dimensions and characteristic areas and volumes of the ship.

```
ENTITY  main_dimension_ratios

  SUBTYPE of (main_dimensions);

(*
DERIVE
  length_breadth_ratio :   REAL  := length_pp/breadth_moulded;
  breadth_draught_ratio :   REAL  :=
```

```
                                  breadth_moulded/draught_moulded;
  length_depth_ratio :   REAL   :=
                               length_pp/depth_to_main_deck;
  breadth_depth_ratio :   REAL   :=
                             breadth_moulded/depth_to_main_deck;
  length_times_breadth :   REAL   :=
                             length_pp * breadth_moulded;
  length_times_breadth_times_depth :   REAL :=
               length_pp breadth_moulded * depth_to_main_deck;
  block_coefficient :   REAL   := displacement_draught_moulded/
                           length_times_breadth_times_depth;
  water_plane_coefficient : REAL   := water_plane_area/
                               (length_wl*breadth_moulded);
  midship_section_coeff : REAL   :=
       midship_section_area/(breadth_moulded*draught_moulded);
  prismatic_coefficient : REAL   :=
                               displacement_draught_moulded/
                               (midship_section_area*length_wl);
  trim_ratio :   REAL   := trim/length_pp;      *)
END_ENTITY;
```

## 37. weight_summary

Collection of information about the principal weight categories of the light ship and related reference quantities for the weight analysis.

```
ENTITY  weight_summary
  SUBTYPE of (general_characteristics);

  steel_weight : metric;
  outfitting_weight : metric;
  machinery_weight : metric;
  electrical_weight : metric;
  spare_weight : metric;
  light_ship_weight : metric;
  frame_spacing_midship_section : metric;
  no_of_slices :  INTEGER;

  (*"slices" of the steel structure as they are used for
     weight estimations, relating to a ship-longitudinal
     segment pattern.                                    *)

  slice_x_centroid : metric;
  slice_weight : metric;
WHERE
  w20:no_of_slices > 0;
END_ENTITY;
```

## 38. machinery

Characteristics of the main propulsion engine and of the auxiliary engines and generators.

```
ENTITY  machinery
  SUBTYPE of (general_characteristics);

  machine_room_location :   STRING;              --aft, amidships
  type_of_main_engine :   SET [1:#] of   STRING;
                               --Diesel, steam turbine etc.
  engine_date_of_build : date;
  cylinder_arrangement : STRING;         --V + degrees, R etc.
```

```
      ignition_type       :  STRING;
      stroke_type         :  STRING;
      stroke_cycle :  STRING;
      bore : metric;
      stroke : metric;
      no_of_main_engines :  INTEGER;
      no_of_cylinders_per_engine :  INTEGER;
      total_power_output_at_MCR : metric;
      power_at_MCR_per_engine :  OPTIONAL  metric;
      max_continuous_rating : metric;
      derated_MCR : metric;
      specific_consumption : metric;
      no_of_gears :  OPTIONAL  INTEGER;
      type_of_gears :  OPTIONAL SET [0:#] of  STRING;
      no_of_auxiliary_engines :  INTEGER;
      type_of_aux_engines :  SET [1:#] of  STRING;
      total_power_output_of_aux : metric;
      no_of_generators :  INTEGER;
      type_of_generators :  SET [1:#] of  STRING;
      total_power_output_of_gens : metric;
      gens_voltage : metric;
      gens_frequency : metric;
      AC_DC_indicator :  STRING;
WHERE
   w21:no_of_main_engines > 0;
   w22:no_of_cylinders_per_engine >= 3;
   w23:no_of_gears >= 0;
   w24:no_of_auxiliary_engines > 0;
   w25:no_of_generators > 0;
END_ENTITY;
```

## 39. propulsion

### Types of propulsion and rudder systems.

```
ENTITY   propulsion
   SUBTYPE of (general_characteristics)
   SUPERTYPE of (propulsion_by_screw OR propulsion_aids);

   type_of_rudder :  STRING;
   type_of_propulsion :  SET [1:#] of  STRING;
END_ENTITY;
```

## 40. propulsion_by_screw

### Subtype of propulsion. Characteristics of the screw propellers.

```
ENTITY propulsion_by_screw
   SUBTYPE of (propulsion);

   no_of_screws :  INTEGER;
   type_of_screw :  SET [0:#] of  STRING;
                           --controllable pitch, high skew etc.
   screw_diameter : metric;
   no_of_blades :  INTEGER;
   screw_rpm :  INTEGER;
   hub_diameter : metric;
   propeller_pitch : metric;
   skew_angle : metric;
   propeller_x_position : metric;
                           --acc. to the ship's coordinate system
   propeller_y_position : metric;
```

```
    propeller_z_position : metric;
    blade_area_projected : metric;
    blade_area_developed : metric;
    blade_area_expanded : metric;
    blade_thickness_at_axis : metric;
DERIVE
    pitch_ratio :  REAL     := propeller_pitch/screw_diameter;
    propeller_disc_area :  REAL     := (screw_diameter**2)*PI/4;
    blade_area_ratio : REAL :=
                          blade_area_expanded/propeller_disc_area;
WHERE
    w26:no_of_screws > 0;
    w27:no_of_blades > 0;
    w28:screw_rpm > 0;
END_ENTITY;
```

## 41. propulsion_aids

**Subtype of propulsion. Characteristics of auxiliary propulsion devices and thrusters.**

```
ENTITY propulsion_aids
    SUBTYPE of (propulsion);

    no_of_propulsion_aids :   INTEGER;        --vane wheel etc.
    type_of_propulsion_aids :  SET  [0:#] of  STRING;
    no_of_thrusters :  INTEGER;
    location_of_thrusters :  SET  [1:#] of  STRING;
    type_of_thrusters :  SET  [0:#] of  STRING; --water jet etc.
    thruster_diameter : metric;
    thrust_of_thruster : metric;
    power_of_thruster : metric;
    thruster_rpm :  INTEGER;
WHERE
    w29:no_of_propulsion_aids >= 0;
    w30:no_of_thrusters >= 0;
    w31:thruster_rpm > 0;
END_ENTITY;
```

## 42. freeboard

**Information related to the regulatory freeboard and associated data based on the freeboard rules.**

```
ENTITY  freeboard
    SUBTYPE of (general_characteristics);

    type_of_freeboard :  STRING;
    length_at_85_percent_depth : metric;
    thickness_of_deckplating : metric;                    --at deck stringer
    depth_for_freeboard : metric;
    block_coefficient_at_85_percent_depth : metric;
    displacement_at_85_percent_depth : metric;
    freeboard : metric;
END_ENTITY;
```

## 43. cargo_and_tank_capacities

**Dry cargo hold and ballast, fuel and lubeoil tank volume capacities.**

```
ENTITY  cargo_and_tank_capacities
    SUBTYPE of (general_characteristics)
```

```
      SUPERTYPE of (insulated_cargo_capacities OR
                    liquid_cargo_capacities OR
                    RORO_cargo_capacities OR
                    container_cargo_capacities);
  no_of_holds :  INTEGER;
  hull_volume : metric;
  upper_works_volume : metric;
  decks_house_volume : metric;
  gross_volume_of_holds : metric;
  grain_volume_of_holds : metric;
  bale_volume_of_holds : metric;
  volume_of_segregated_BW_tanks : metric;
  volume_of_clean_BW_tanks : metric;
  volume_of_FO_tanks : metric;
  heating_coils_indicator :  STRING;
  volume_of_LO_tanks : metric;
  volume_of_FW_tanks : metric;
WHERE
  w32:no_of_holds >= 1;
END_ENTITY    ;
```

## 44. insulated_cargo_capacities

### Volume capacity of insulated holds.

```
ENTITY insulated_cargo_capacities
  SUBTYPE of (cargo_and_tank_capacities);

  volume_of_insulated_holds : metric;
END_ENTITY;
```

## 45. liquid_cargo_capacities

### Volume capacity of liquid cargo tanks.

```
ENTITY liquid_cargo_capacities
  SUBTYPE of (cargo_and_tank_capacities);

  volume_of_liquid_cargo_holds : metric;
  volume_of_liquid_gas_holds : metric;
END_ENTITY;
```

## 46. RORO_cargo_capacities

### Capacities for RORO vehicles.

```
ENTITY RORO_cargo_capacities
  SUBTYPE of (cargo_and_tank_capacities);

  no_of_RORO_lanes :  INTEGER;
  length_of_RORO_lanes : metric;
  clear_height_of_RORO_lanes : metric;
WHERE
  w33:no_of_RORO_lanes >= 0;
END_ENTITY;
```

## 47. container_cargo-capacities

Capacity for container cargo on deck and in holds, connection plugs for refrigerated containers.

```
ENTITY container_cargo_capacities
   SUBTYPE of (cargo_and_tank_capacities);

   no_of_on_deck_TEU_lots :   INTEGER;
   no_of_on_deck_FEU_lots :   INTEGER;
   no_of_TEU_lots_in_hold :   INTEGER;
   no_of_FEU_lots_in_hold :   INTEGER;
   TEU_total_capacity :   INTEGER;
   no_of_reefer_plugs :   OPTIONAL INTEGER;

   (* TEU = Twenty Foot Equivalent Unit
      FEU = Forty Foot Equivalent Unit *)

WHERE
   w34:no_of_on_deck_TEU_lots >= 0;
   w35:no_of_on_deck_FEU_lots >= 0;
   w36:no_of_TEU_lots_in_hold >= 0;
   w37:no_of_FEU_lots_in_hold >= 0;
   w38:no_of_reefer_plugs >= 0;
   w39:TEU_total_capacity >= 0;
   w40:no_of_reefer_plugs < TEU_total_capacity;
END_ENTITY;
```

## 48. crew_and_passenger_capacities

**Number of crew, drivers, and passengers. Number and type of accommodation.**

```
ENTITY    crew_and_passenger_capacities
    SUBTYPE of (general_characteristics)
    SUPERTYPE of (cruise_passenger_capacities);

   no_of_crew :  INTEGER;
   no_of_drivers_berthed :  OPTIONAL INTEGER;
   no_of_cabins :  INTEGER;
   no_of_cabin_types :  INTEGER;
   no_of_each_cabin_type    :  SET [1:#] of INTEGER;
   type_of_cabins :   SET [1:#] of  STRING;
   size_of_cabin : metric;
   no_of_passengers :  INTEGER;
WHERE
   w41:no_of_crew >= 1;
   w42:no_of_drivers_berthed >= 0;
   w43:no_of_cabins >= 1;
   w44:no_of_cabin_types >= 1;
   w45:no_of_each_cabin_type  >= 1;
   w46:no_of_passengers >= 0;
END_ENTITY;
```

## 49. cruise_passenger_capacities

**Capacities for cruise passengers.**

```
ENTITY  cruise_passenger_capacities
   SUBTYPE of (crew_and_passenger_capacities);

   no_of_cruise_passengers :   INTEGER;
   no_of_passenger_cabins :   INTEGER;
   no_of_passenger_cabin_types :   INTEGER;
   hotel_area : metric;
   public_area : metric;
   service_area : metric;
   outdoor_area : metric;
```

```
  WHERE
    w47:no_of_cruise_passengers > 0;
    w48:no_of_passenger_cabins >= 0;
    w49:no_of_passenger_cabin_types >= 0;
END_ENTITY;
```

## 50. cargo_access_and_auxiliary_equipment

Types and capacities of cargohandling equipment, hatch closures, deck auxiliaries, rescue and nautical systems, board computers.

```
ENTITY  cargo_access_and_auxiliary_eqmt
  SUBTYPE of (general_characteristics)
  SUPERTYPE of (RORO_equipment);

  (* CAAE = cargo access and auxiliary equipment *)

  no_of_CAAE_items :  INTEGER;
  type_of_CAAE_items :  SET [1:#] of  STRING;
  no_of_cranes :  OPTIONAL INTEGER;
  type_of_cranes :  OPTIONAL SET [0:#] of  STRING;
                                      --e.g. masthead crane
  SWL_of_cranes :  OPTIONAL LIST [0:#] of metric;
                                      --e.g. 40 t, 20 t, 1,5 t
  no_of_hatch_covers :  INTEGER;
  type_of_hatch_covers :  SET [1:#] of  STRING;
                                      --pontoon, piggy back etc.
  hatch_dimensions : SET [0:#] of SET [2:2] of REAL;
                                      --e.g. 3x3, 5x5 (metres) etc.
  no_of_deck_machines :  INTEGER;
  type_of_deck_machines :  SET  [1:#] of STRING;
  no_of_rescue_devices :  INTEGER;
  type_of_rescue_devices :  SET  [1:#] of STRING;
                                      --e.g. closed dinghies
  no_of_nautical_devices :  INTEGER;
  type_of_nautical_devices :  SET [1:#] of STRING;
                                      --e.g. LORAN, SATNAV, GPS
  no_of_board_computers :  OPTIONAL INTEGER;
  type_of_board_computers :  OPTIONAL SET [0:#] of  STRING;
WHERE
  w50:no_of_board_computers >= 0;
  w51:no_of_CAAE_items > 0;
  w52:no_of_cranes >= 0;
  w53:no_of_hatch_covers > 0;
  w54:no_of_deck_machines > 0;
  w55:no_of_rescue_devices > 0;
  w56:no_of_nautical_devices > 0;
END_ENTITY;
```

## 51. RORO_equipment

Types and capacities of RORO vehicle handling equipment.

```
ENTITY  RORO_equipment
  SUBTYPE of (cargo_access_and_auxiliary_eqmt);

  no_of_doors :  INTEGER;
  type_of_doors :  SET [0:#] of  STRING;
  position_of_doors :  SET [0:#] of  STRING;
                                      --bow, side, stern
  no_of_RORO_ramps :  INTEGER;
  position_of_RORO_ramps :  SET [0:#] of  STRING;
```

```
  SWL_of_RORO_ramps :  SET  [0:#] of  STRING;
  axle_weight_max_per_ramp :  SET  [0:#] of  STRING;
WHERE
  w57:no_of_RORO_ramps >= 0;
  w58:no_of_doors >= 0;
END_ENTITY;
```

## 52. factors

### Fuel oil and lube oil density factors.

```
ENTITY  factors
  SUBTYPE of (general_characteristics);

  FO_density : metric;
  LO_density : metric;
END_ENTITY;
```

## 53. physical_constants

Not legal EXPRESS at the time of the model, hence marked as a comment here. But now valid EXPRESS by declaration of constants.

```
(*
  CONSTANT

  g_const :  REAL  := 9.815 ;
  elastic_modulus_steel : NUMBER := 210000;
  END_CONSTANT;
*)
```

## 54. delivery_information

### Information to ship owner at delivery time.

```
ENTITY  delivery_information
  SUBTYPE of (general_characteristics);

  built_at_yard :  STRING;
  building_no :  STRING;
  ships_name :  STRING;
  former_names :  OPTIONAL SET  [0:#] of  STRING;
  radio_call_signal :  STRING;
  sat_communication_id :  STRING;
  port_of_registry :  STRING;
  owner :  STRING;
  managing_company :  STRING;
  date_of_contract : date;
  date_of_laying_down : date;
  date_of_launching : date;
  date_of_delivery : date;
UNIQUE
  u3:date_of_contract, date_of_laying_down,
     date_of_launching, date_of_delivery;
  u4:ships_name, radio_call_signal,
     sat_communication_id,  port_of_registry, building_no;
WHERE
  w60:date_of_contract < date_of_laying_down;
  w61:date_of_laying_down < date_of_launching;
  w62:date_of_launching < date_of_delivery;
END_ENTITY;
```

## 55. performance

Collection of physical and economic performance criteria for ship.

```
ENTITY  performance
  SUPERTYPE of  (ONEOF (ballast_performance,
                       cargo_transport_performance));

  metacentric_radius : metric;
  metacentric_height : metric;

(*
  gravity : CONSTANT;
*)

  FN_input : main_dimensions;
  light_ship_x_centroid : metric;
                              --acc. to ship's coordinate system
  light_ship_y_centroid : metric;
  light_ship_z_centroid : metric;
  required_freight_rate : metric;
  ship_merit_factor : metric;
  sand_roughness : metric;

(*
DERIVE
  Froude_number :  REAL    := FN_input.speed_operating/
                    SQRT(FN_input.length_wl*gravity.g_const);
*)

END_ENTITY;
```

## 56. ballast_performance

Physical performance data in ballast condition.

```
ENTITY  ballast_performance
  SUBTYPE of (performance);

  b_stability_max_lever : metric;
  b_range_of_stability : metric;                      --degrees
  b_stability_max_righting_moment : metric;
  b_max_free_surfaces_area  : metric;
  b_max_bending_moment_still_water : metric;
  b_max_bending_moment_waves : metric;
  b_total_max_bending_moment : metric;
  b_max_local_structural_loads : SET [0:#] of STRING;
                                    --wheeled cargo etc.
  b_total_resistance_acc_to : SET [1:#] of STRING;
                              --prediction method e.g. ITTC
  b_resistance_coeff_friction : metric;
  b_resistance_coeff_pressure : metric;
  b_resistance_coeff_wavemaking : metric;
  b_resistance_coeff_air : metric;
  b_resistance_coeff_sea_motion : metric;
  b_resistance_coeff_spray : metric;
  b_resistance_coeff_total_ice :  OPTIONAL  metric;
  b_motion_period_heave : metric;
  b_motion_period_roll : metric;
  b_motion_period_pitch : metric;
  b_speed_power_equation :  STRING;
END_ENTITY;
```

## 57. cargo_transport_performance

**Subtype of performance, supertype of performance at beginning and end of trip.**

```
ENTITY  cargo_transport_performance
  SUBTYPE of (performance)
  SUPERTYPE of (ONEOF(ctp_begin_of_trip,
                      ctp_end_of_trip));

END_ENTITY;
```

## 58. ctp_begin_of_trip

**Performance data for beginning of trip in full load.**

```
ENTITY ctp_begin_of_trip

  (* ctp stands for cargo_transport_performance, the "b"
     indicates begin_of_trip, the "c" indicates the cargo
     load case.                                            *)

  SUBTYPE of (cargo_transport_performance);

  b_cargo_x_centroid_homogeneous : metric;
                            --acc. to the ship's coordinates
  b_cargo_y_centroid_homogeneous : metric;
  b_cargo_z_centroid_homogeneous : metric;
  b_cargo_x_centroid_per_hold :  SET [1:#]   of  metric;
  b_cargo_y_centroid_per_hold :  SET [1:#]   of  metric;
  b_cargo_z_centroid_per_hold :  SET [1:#]   of  metric;
  b_c_stability_max_lever : metric;
  b_c_range_of_stability : metric;
  b_c_stability_max_righting_moment : metric;
  b_c_max_free_surfaces_area  : metric;
  b_c_max_bending_moment_still_water : metric;
  b_c_max_bending_moment_waves : metric;
  b_c_total_max_bending_moment : metric;
  b_c_max_local_structural_loads :  SET  [0:#]  of  STRING;
  b_c_total_resistance_acc_to :  SET [1:#] of  STRING;
  b_c_resistance_coeff_friction : metric;
  b_c_resistance_coeff_pressure : metric;
  b_c_resistance_coeff_wavemaking : metric;
  b_c_resistance_coeff_air : metric;
  b_c_resistance_coeff_sea_motion : metric;
  b_c_resistance_coeff_spray : metric;
  b_c_resistance_coeff_total_ice :  OPTIONAL metric;
  b_c_motion_period_heave : metric;
  b_c_motion_period_roll : metric;
  b_c_motion_period_pitch : metric;
  b_c_speed_power_equation :  STRING;
END_ENTITY;
```

## 59. ctp_end_of_trip

**Performance data for end of trip in full load.**

```
ENTITY ctp_end_of_trip

  (* ctp stands for cargo_transport_performance, the "e"
     indicates end_of_trip, the "c" indicates the cargo load
     case.                                                  *)
  SUBTYPE of (cargo_transport_performance);
```

```
   e_cargo_x_centroid_homogeneous : metric;
                              --acc. to the ship's coordinates
   e_cargo_y_centroid_homogeneous : metric;
   e_cargo_z_centroid_homogeneous : metric;
   e_cargo_x_centroid_per_hold : SET [1:#] of  metric;
   e_cargo_y_centroid_per_hold : SET [1:#] of  metric;
   e_cargo_z_centroid_per_hold : SET [1:#] of  metric;
   e_c_stability_max_lever : metric;
   e_c_range_of_stability : metric;
   e_c_stability_max_righting_moment : metric;
   e_c_max_free_surfaces_area  : metric;
   e_c_max_bending_moment_still_water : metric;
   e_c_max_bending_moment_waves : metric;
   e_c_total_max_bending_moment : metric;
   e_c_max_local_structural_loads :  SET  [0:#] of STRING;
   e_c_total_resistance_acc_to :  SET [1:#] of STRING;
   e_c_resistance_coeff_friction : metric;
   e_c_resistance_coeff_pressure : metric;
   e_c_resistance_coeff_wavemaking : metric;
   e_c_resistance_coeff_air : metric;
   e_c_resistance_coeff_sea_motion : metric;
   e_c_resistance_coeff_spray : metric;
   e_c_resistance_coeff_total_ice : OPTIONAL metric;
   e_c_motion_period_heave : metric;
   e_c_motion_period_roll : metric;
   e_c_motion_period_pitch : metric;
   e_c_speed_power_equation :  STRING;
END_ENTITY;
```

## 60. rationale

Supertype of design context information.

```
ENTITY rationale
   SUPERTYPE of (ONEOF (environmental_data, standards,
                        design_principles));
END_ENTITY;
```

## 61. environmental_data

Environmental information about the sea water, seaway, and air.

```
ENTITY  environmental_data
  SUBTYPE of (rationale);

(*
  gravity : CONSTANT;
*)

  sea_water_density : metric;
  sea_water_temp : metric;
  sea_water_viscosity : metric;
  max_wave_height : metric;
  max_wave_length : metric;
  max_wave_slope :  INTEGER;                    --in degrees
  average_wave_height : metric;
  average_wave_length : metric;
  max_air_temp : metric;
  min_air_temp : metric;
  max_air_humidity :  INTEGER;

(*
```

```
DERIVE
  max_wave_period : REAL :=
                    SQRT(2*PI*max_wave_length/gravity.g_const);
  average_wave_period :   REAL    :=
              SQRT(2*PI*average_wave_length/gravity.g_const);
*)

END_ENTITY;
```

### 62. standards

**Applicable standards as design reference.**

```
ENTITY standards
  SUBTYPE of (rationale);

  attributes :  STRING;                      --to be defined
END_ENTITY;
```

### 63. design_principles

**Methods and strategies of design.**

```
ENTITY  design_principles
  SUBTYPE of (rationale);

  more_attributes :  STRING;                 --to be defined
END_ENTITY;
END_SCHEMA;
```

## 2  EXPRESS Representations of the Ship Design and Production Engineering Entities

Proposed EXPRESS codes corresponding to the same models included in Annex V are included here. They are:

- Ship Design Definition (see Fig. 67)
- Ship Design Product (see Fig. 68)
- Ship Design Structural Element (see Fig. 69)
- Ship Production Engineeriny Product (see Fig. 70)
- Ship Production Engineering Prefabricated Block (see Fig. 71)

```
(* Design_Definition *)
ENTITY Design_Definition
SUBTYPE OF (Definition);
SUPERTYPE OF (ONEOF(Design_Product, Design_Activity));
with_design_type: design_type;
identified_by: design_code;
UNIQUE
identified_by;
END ENTITY;

TYPE
design_type = ENUMERATION OF (product, activity);
design_code = STRING;
END_TYPE;
```

```
(* D_Design_Product *)
ENTITY Design_Product
SUBTYPE OF (Design_Definition, Product);
SUPERTYPE OF ( ONEOF (D_Plate, D_Commercial_Stiffener,
   D_Connecting_Part, D_Flange, D_Web, D_Hull,
   D_Structural_Element, D_Primary_Substructure,
   D_Plate_Sheet, D_Plate_Assembly, D_Composite_Stiffener,
   D_Opening, D_Connection, D_Joint, D_Space,
   D_Compartment)
   XOR NULL),
with_design_product_type: design_product_type; technical_soultion_of:
Function;
with_design_cost: Cost;
WHERE
'PRODUCT_FUNCTION' IN TYPEOF (technical_solution_of_function);
END_ENTITY

TYPE
design_product_type = ENUMERATION OF (plate, commercial stiffener,
connecting_part, flange, web, hull, structural_element,
primery_suhstructure, plate_sheet, plate_assembly, composite_stiffener,
opening, connection, joint, space, compartment);
END_TYPE;
```

```
(* D_Structural_Element *)
ENTITY D_Structural_Element
SUBTYPE OF (Design_Product, Assembly_Part);
with_structural_element_type: structural_element_type;
lies_on: SET il:P~C OF D_Moulded_Surface;
connected_to_design_product_by: LIST jl:PtC OF D_Connecting_Part;
WHERE
'D_HULL' IN TYPEOF (component_of_parent_part);
'D_PRIMARY_SUBSTRUCTURE? IN TYPEOF (composed_of);
END ENTITY;
TYPE
structural_element_type - ENUMERATION OF (double_bottom,
                         bulkhead, deck);
END_TYPE;
```

```
(* Production_Engineering_Product *)
ENTITY P_E_Product
SUBTYPE OF (Production_Engineering_Definition, Product);
SUPERTYPE OF (ONEOF( P_E_Product_Plate, P_E_Product_
   Commercial_Stiffener, P_E_Product_Connecting_Part,
   P_E_Product_Flange, P_E_Product_Web, P_E_Product_Hull,
   P_E_Product_Sub_Assembly, P_E_Product_Assembly,
   P_E_Product_Prefabricated_Block,
   P_E_Product_Prefabricated_Sub_Block,
   P_E_Product_Opening, P_E_Product_Composite_Stiffener,
   P_E_Product_Connection, P_E_Product_Joint,
   P_E_Product_Compartment)
   XOR NULL);
with_production_enginering_product_type: p_e_product_type;
with_estimating_material_cost: SET iO:1c OF Cost;
with_estimating_cost: Cost;
END_ENTITY;

TYPE
p_e_product_type =    ENUMERATION OF (plate, commercial_
stiffener, connecting_part, flange,
web, hull, sub_assembly, assembly,
prefabricated_block,
prefabricated_sub_block, opening,
```

```
composite_stiffener, connection,
joint, compartment)
END TYPE;
```

**(* Production_Engineering_Prefabricated_Block *)**
```
ENTITY P_E_Prefabricated_Block
SUBTYPE OF (P_E_Product, Assembly_Part),
planned_product_of: P_Prefabricated_Block;
cut_by: LIST jO:PtC OF P_E~Opening;
penetrated_by: LIST jO:Pt~ OF P_E_Cutout_Hole;
stiffened_by: LIST j1:Pt~ OF P_E_Stiffener;
connected_to_design_product_by: LIST j1:Pt~ OF P_E_Connecting_Part;
joined_to_design_product_by: LIST j1:Pt~ OF UNI~UE P_E_Joint;
with_thickness: Length_Measure;
with_thickness_orientation: thickness_orientation;
WHERE
'P_E_HULL' IN TYPEOF (component_of_parent_part);
'P_E_SU8BLOCK' IN TYPEOF (composed_of);
OR
'P_E_STIFFENING_ELEMENT_ASSEMBLY' IN TYPEOF (composed_of);
END_ENTITY;
```

## List of Figures

# Springer-Verlag
# and the Environment

We at Springer-Verlag firmly believe that an international science publisher has a special obligation to the environment, and our corporate policies consistently reflect this conviction.

We also expect our business partners – paper mills, printers, packaging manufacturers, etc. – to commit themselves to using environmentally friendly materials and production processes.

The paper in this book is made from low- or no-chlorine pulp and is acid free, in conformance with international standards for paper permanency.

## Research Reports ESPRIT

### Area *Computer-Integrated Manufacturing and Engineering (CIME)*

Improving the Performance of Neutral File Data Transfers. Edited by R.J. Goult, P.A. Sherar. IX, 138 pages. 1990 (Project 322 CAD*I, CAD Interfaces, Vol. 6)

Advanced Modelling for CAD/CAM Systems. Edited by H. Grabowski, R. Anderl, M.J. Pratt. VI, 113 pages. 1991 (Project 322 CAD*I, Vol. 7)

IMPPACT Reference Model. Edited by W.F. Gielingh, A.K. Suhm. XII, 261 pages. 1993 (Project 2165 IMPPACT, Integrated Modelling of Products and Processes using Advanced Computer Technologies)

CIMOSA: Open System Architecture for CIM. Edited by ESPRIT Consortium AMICE. XI, 234 pages. 2nd, rev. and ext. edition 1993 (Project 688/5288 AMICE, A European CIM Architecture)

Vibration Control of Flexible Servo Mechanisms. Edited by J.-L. Faillot. VII, 206 pages, 1993 (Project 1561 SACODY, A High Performance Flexible Manufacturing System (FMS) Robot with On-Line Dynamic Compensation)

Neutral Interfaces in Design, Simulation, and Programming or Robotics. Edited by I. Bey et al. XV, 334 pages, 6 figs. 1994 (Project 2614/5105 NIRO, Neutral Interfaces for Robotics)

CCE: An Integration Platform for Distributed Manufacturing Applications. A Survey of Advanced Computing Technologies. Edited by ESPRIT Consortium CCE-CNMA. XII, 207 pages. 1995 (Project 7096 CCE-CNMA, CIME Computing Environment: Integrating CNMA, Vol. 1)

MMS: A Communication Language of Manufacturing. Edited by ESPRIT Consortium CCE-CNMA. XII, 185 pages. 1995 (Project 7096 CCE-CNMA, Vol. 2)

### Subseries PDT (Product Data Technology)

CAD Geometry Data Exchange Using STEP. Edited by H.J. Helpenstein. XIV, 432 pages. 1993 (Project 2195 CADEX, CAD Geometry Data Exchange)

NEUTRABAS. A Neutral Product Definition Database for Large Multifunctional Systems. Edited by H. Nowacki. XII, 203 pages. 1995 (Project 2010 NEUTRABAS)